A. R. CLAPHAM T. G. TUTIN E. F. WARBURG

FLORA OF THE BRITISH ISLES

ILLUSTRATIONS

PART II
ROSACEAE–POLEMONIACEAE

DRAWINGS BY
SYBIL J. ROLES

CAMBRIDGE
AT THE UNIVERSITY PRESS
1960

PUBLISHED BY

THE SYNDICS OF THE CAMBRIDGE UNIVERSITY PRESS

Bentley House, 200 Euston Road, London, N.W. 1
American Branch: 32 East 57th Street, New York 22, N.Y.

©

CAMBRIDGE UNIVERSITY PRESS
1960

Printed in Great Britain at the University Press, Cambridge
(Brooke Crutchley, University Printer)

PREFACE

The second part of these Illustrations of British plants follows the same general pattern as the first.

The intention once again is to provide a visual impression of the habit, and a selection of the chief features of the plants so as to assist with the appreciation of the technical descriptions given in *Flora of the British Isles* and the *Excursion Flora*.

The drawings have almost all been made from fresh specimens and are reproduced on a rather larger scale than that of the illustrations accompanying Bentham and Hooker's well-known *Handbook*. It is hoped that these features, combined with the grouping of illustrations to facilitate as far as possible easy comparison of related species will make them generally useful.

We should like to express once again our gratitude to Miss Roles for her patience and skill in the almost impossible task of providing small-scale drawings of 'typical' specimens.

We are greatly indebted to the following for supplying many of the plants illustrated in this Part: Miss J. Allison, Miss M. Atkins, P. W. Ball, P. R. Bell, M. Borrill, Miss A. P. Conolly, R. W. David, Miss E. W. Davies, Miss U. K. Duncan, J. R. S. Fincham, H. Gilbert-Carter, the late R. A. Graham, G. Halliday, M. K. Hanson, Miss J. M. Hartshorn, Miss J. E. Hibberd, E. K. Horwood, H. M. Hurst, A. C. Jermy, J. E. Lousley, D. McClintock, Miss P. A. Padmore, C. D. Piggot, M. E. D. Poore, T. E. D. Poore, C. E. Raven, J. E. Raven, N. W. Simmons, F. A. Sowter, Mrs F. le Sueur, F. J. Taylor, Mrs F. J. Taylor, Miss E. M. Thomas, D. H. Valentine, S. M. Walters, Miss M. McCullum Webster, W. T. Williams, and P. F. Yeo.

A.R.C.
T.G.T.
E.F.W.

PART II

ROSACEAE—POLEMONIACEAE

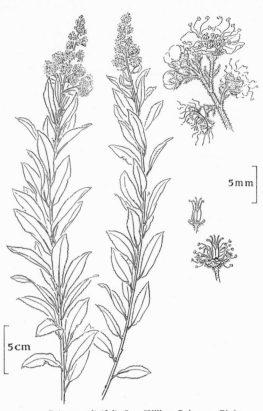

553. *Spiraea salicifolia* L.   Willow Spiraea   Pink

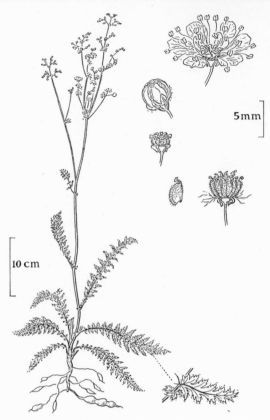

554. *Filipendula vulgaris* Moench   Dropwort   White

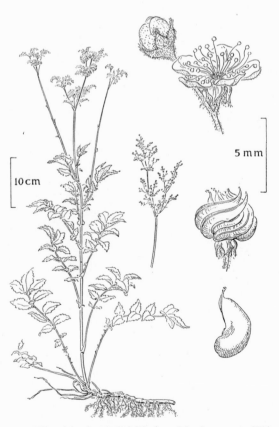

555. *Filipendula ulmaria* (L.) Maxim.   Meadow-sweet   White

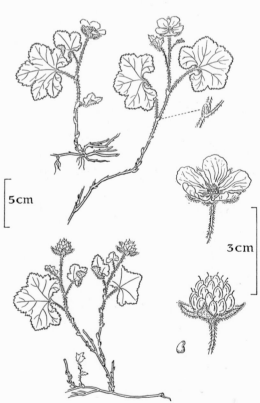

556. *Rubus chamaemorus* L.   Cloudberry   White

557. *Rubus saxatilis* L.   Stone Bramble   White

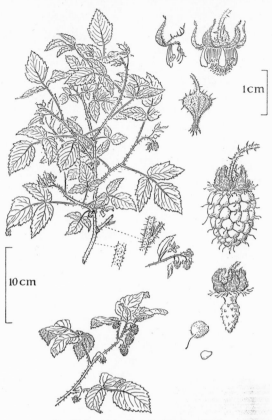

558. *Rubus idaeus* L.   Raspberry   White

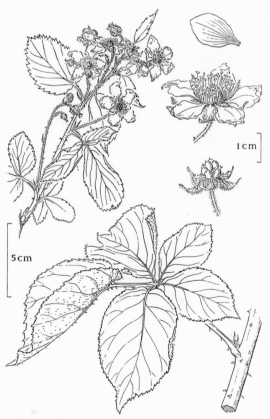

559. *Rubus nessensis* W. Hall   Blackberry   White

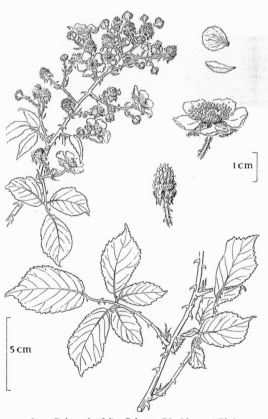

560. *Rubus ulmifolius* Schott   Blackberry   Pink

[3]

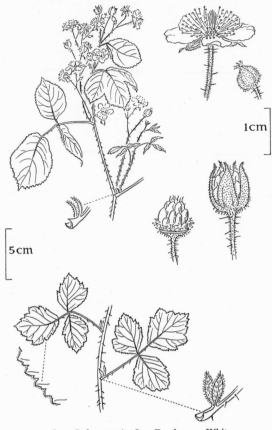

561. *Rubus caesius* L.   Dewberry   White

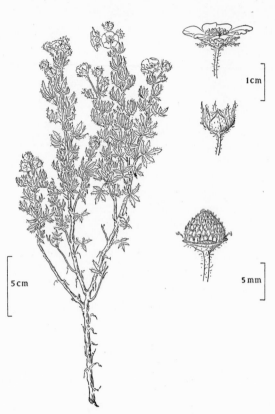

562. *Potentilla fruticosa* L.   Shrubby Cinquefoil
Yellow

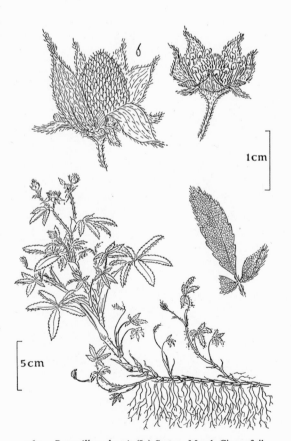

563. *Potentilla palustris* (L.) Scop.   Marsh Cinquefoil
Purple

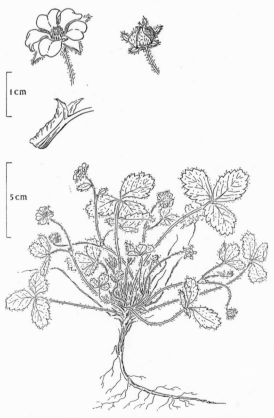

564. *Potentilla sterilis* (L.) Garcke   Barren Strawberry
White

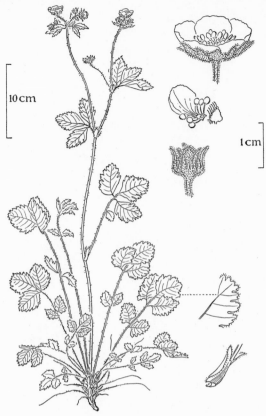

565. *Potentilla rupestris* L.   Rock Cinquefoil   White

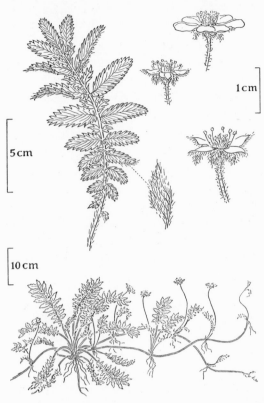

566. *Potentilla anserina* L.   Silverweed   Yellow

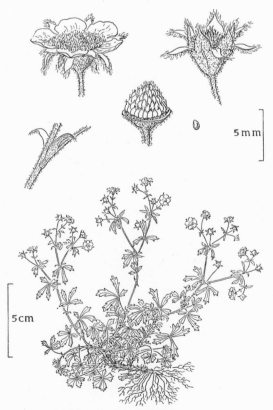

567. *Potentilla argentea* L.   Hoary Cinquefoil   Yellow

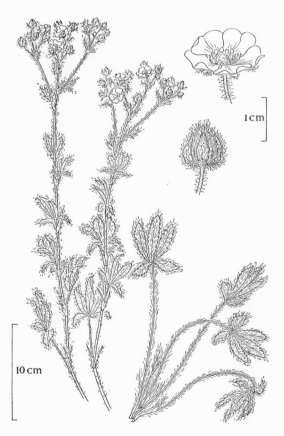

568. *Potentilla recta* L.   Yellow

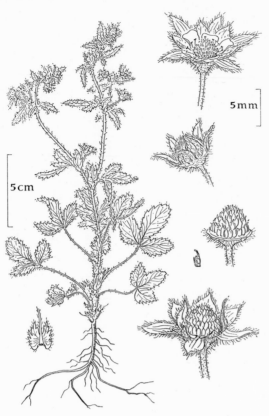

569. *Potentilla norvegica* L.   Yellow

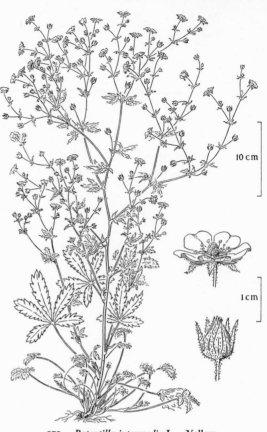

570. *Potentilla intermedia* L.   Yellow

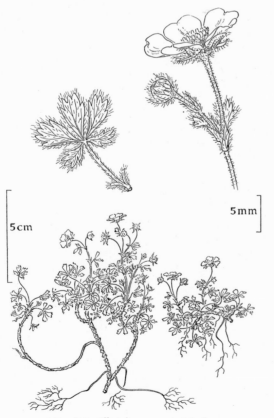

571. *Potentilla tabernaemontani* Aschers.
Spring Cinquefoil   Yellow

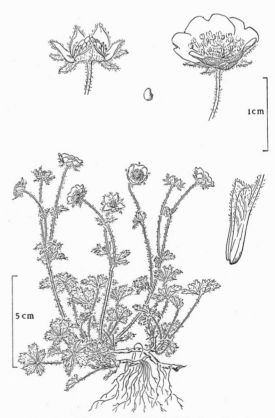

572. *Potentilla crantzii* (Crantz) G. Beck ex Fritsch
Alpine Cinquefoil   Yellow

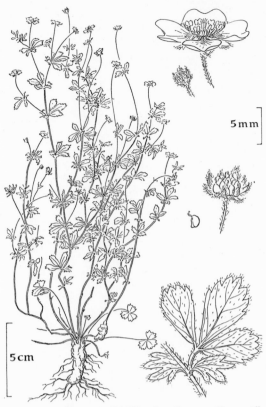

573. *Potentilla erecta* (L.) Räusch.   Common Tormentil
Yellow

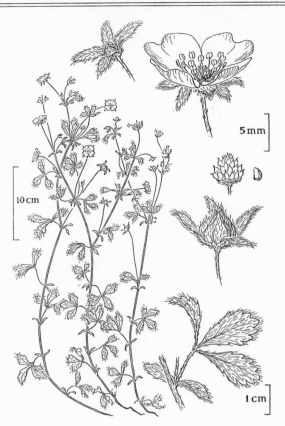

574. *Potentilla anglica* Laicharding   Trailing Tormentil
Yellow

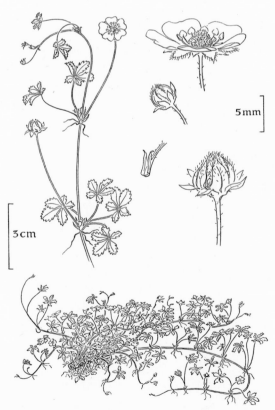

575. *Potentilla reptans* L.   Creeping Cinquefoil   Yellow

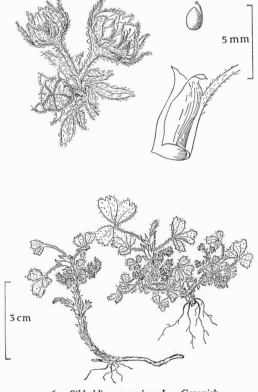

576. *Sibbaldia procumbens* L.   Greenish

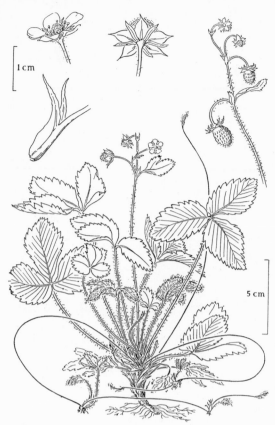

577. *Fragaria vesca* L.   Wild Strawberry   White

578. *Fragaria moschata* Duchesne
Hautbois Strawberry   White

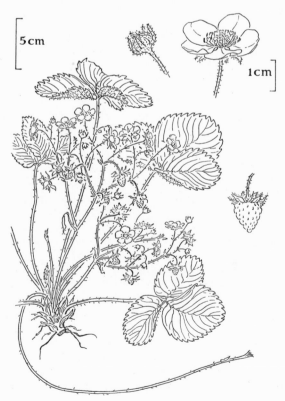

579. *Fragaria ananassa* Duchesne   Garden Strawberry
White

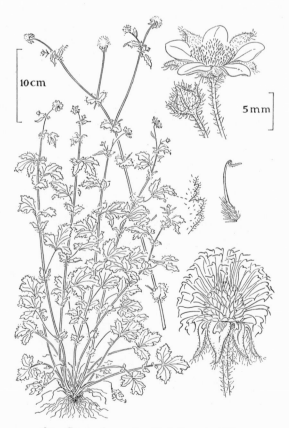

580. *Geum urbanum* L.   Herb Bennet   Yellow

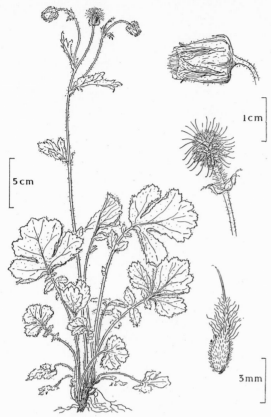

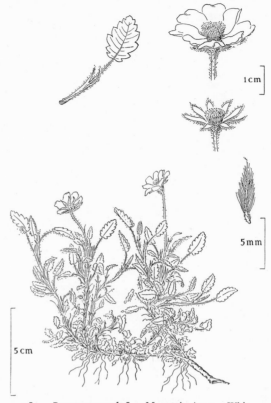

581. *Geum rivale* L.   Water Avens   Orange-pink

582. *Dryas octopetala* L.   Mountain Avens   White

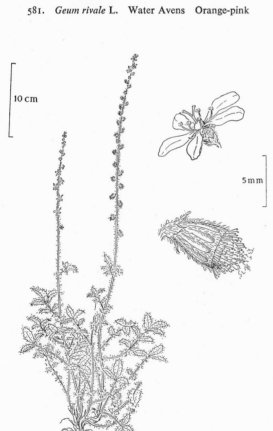

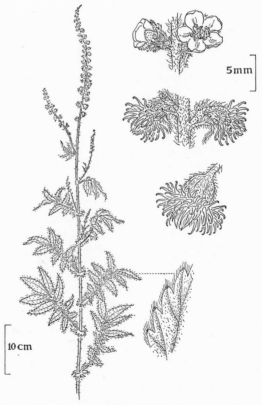

583. *Agrimonia eupatoria* L.   Common Agrimony   Yellow

584. *Agrimonia odorata* (Gouan) Mill.   Fragrant Agrimony   Yellow

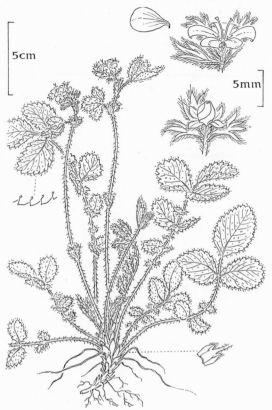

585. *Aremonia agrimonoides* (L.) DC. Yellow

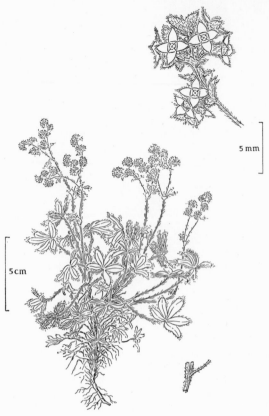

586. *Alchemilla alpina* L. Alpine Lady's-Mantle Greenish

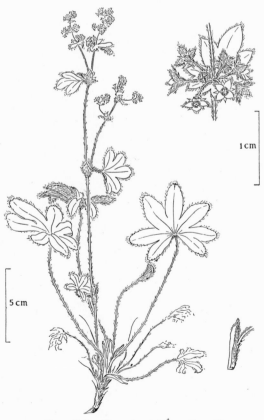

587. *Alchemilla conjuncta* Bab. Greenish

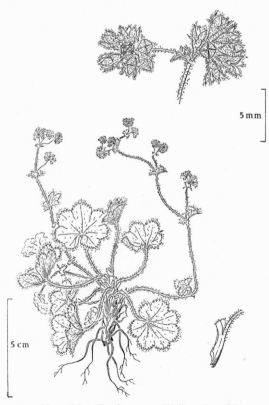

588. *Alchemilla glaucescens* Wallr. Greenish

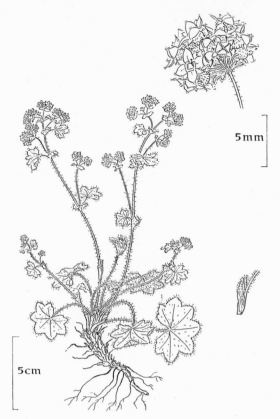

589. *Alchemilla vestita* (Buser) Raunk.   Greenish

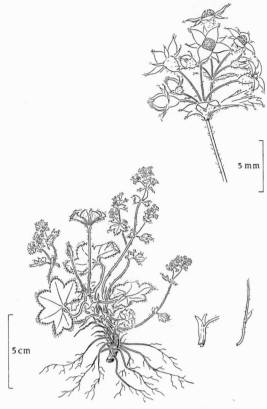

590. *Alchemilla filicaulis* Buser   Greenish

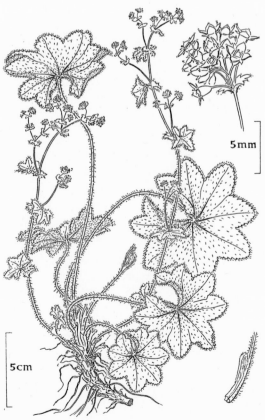

591. *Alchemilla subcrenata* Buser   Greenish

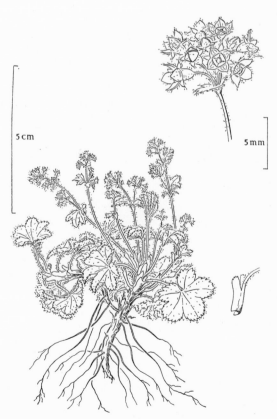

592. *Alchemilla minima* Walters   Greenish

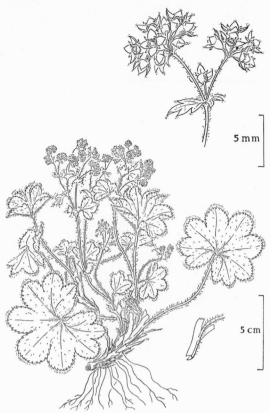

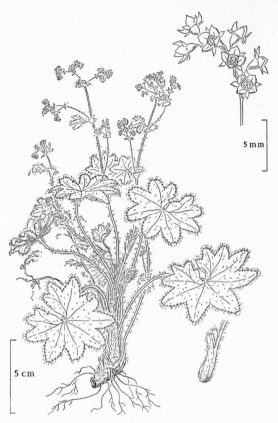

593. *Alchemilla monticola* Opiz   Greenish

594. *Alchemilla acutiloba* Opiz   Greenish

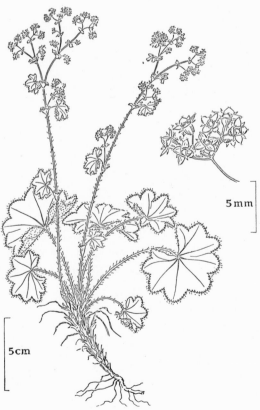

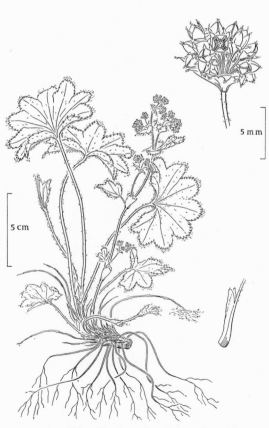

595. *Alchemilla xanthochlora* Rothm.   Greenish

596. *Alchemilla glomerulans* Buser   Greenish

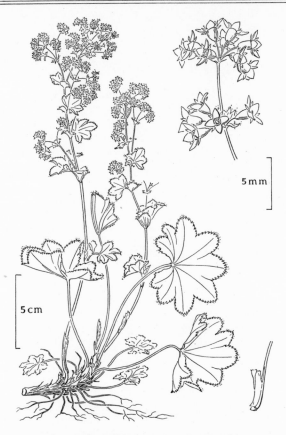

597. *Alchemilla glabra* Neygenf.   Greenish

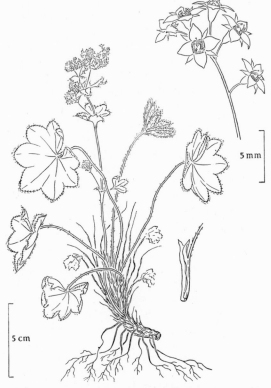

598. *Alchemilla wichurae* (Buser) Stefánsson   Greenish

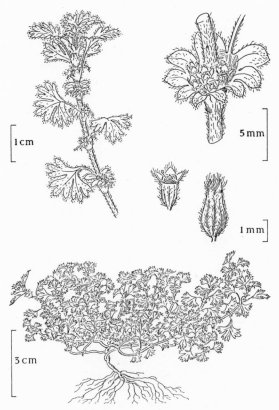

599. *Aphanes arvensis* L.   Parsley Piert   Greenish

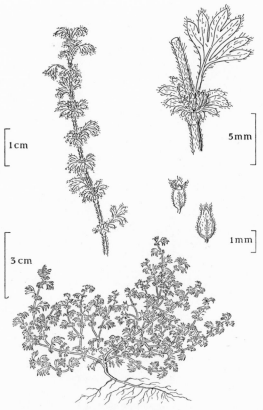

600. *Aphanes microcarpa* (Boiss. & Reut.) Rothm.   Greenish

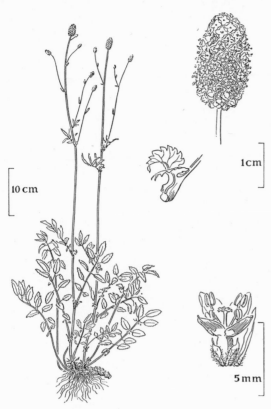

601. *Sanguisorba officinalis* L.   Great Burnet   Crimson

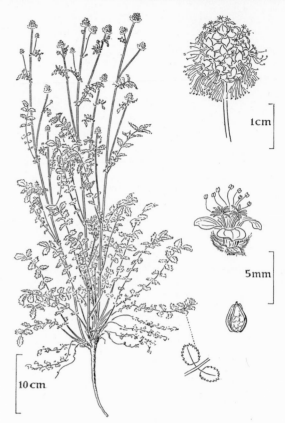

602. *Poterium sanguisorba* L.   Salad Burnet   Greenish

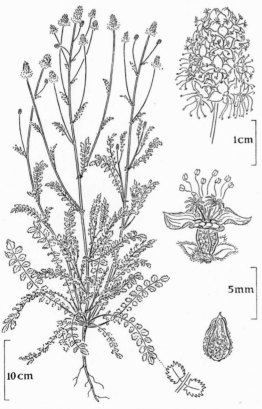

603. *Poterium polygamum* Waldst. & Kit.   Greenish

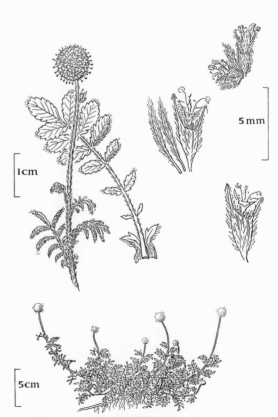

604. *Acaena anserinifolia* (J. R. & G. Forst.) Druce
Greenish

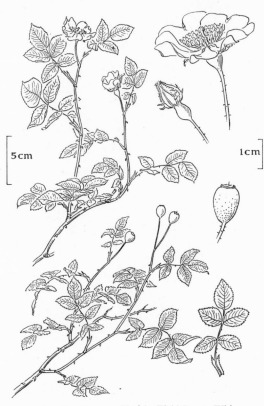

605. *Rosa arvensis* Huds.   Field Rose   White

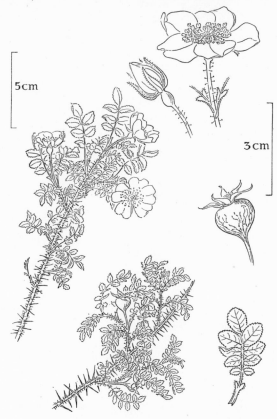

606. *Rosa pimpinellifolia* L.   Burnet Rose   White

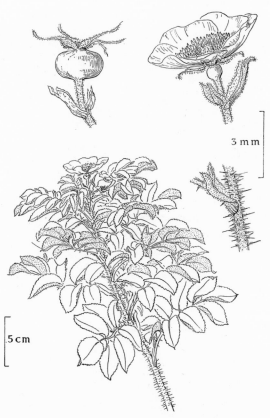

607. *Rosa rugosa* Thunb.   Pink or white

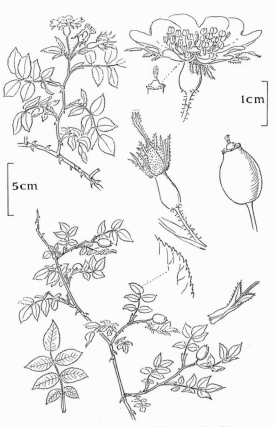

608. *Rosa stylosa* Desv.   White or pale pink

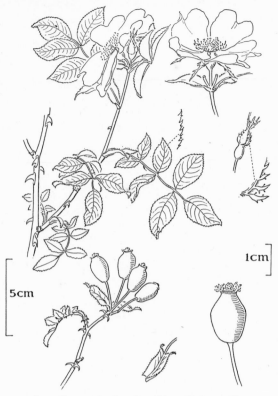

609. *Rosa canina* L.   Dog Rose   Pink or white

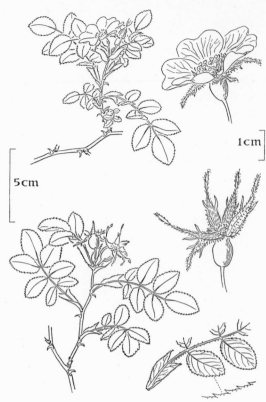

610. *Rosa dumalis* Bechst.   Pink or white

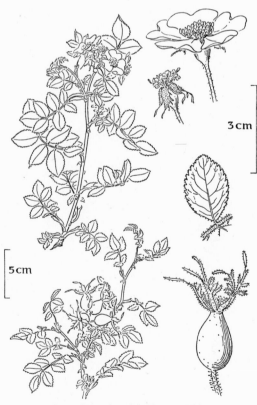

611. *Rosa obtusifolia* Desv.   White

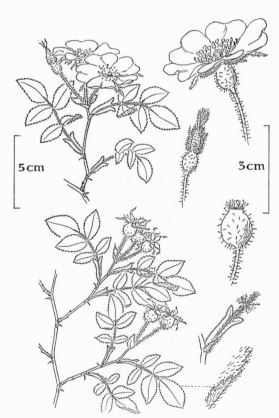

612. *Rosa tomentosa* Sm.   Pink or white

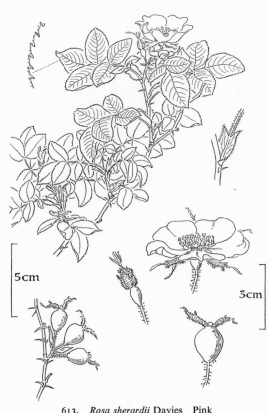

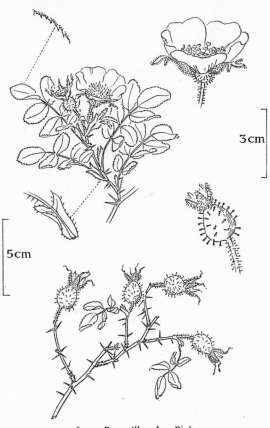

613. *Rosa sherardii* Davies   Pink

614. *Rosa villosa* L.   Pink

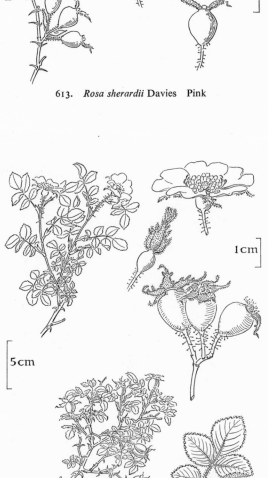

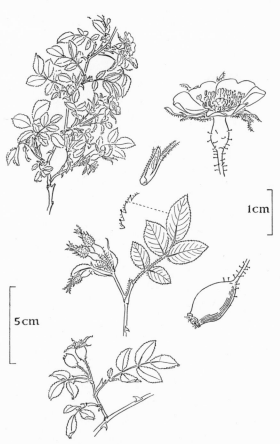

615. *Rosa rubiginosa* L.   Sweet-briar   Pink

616. *Rosa micrantha* Borrer ex Sm.   Pink

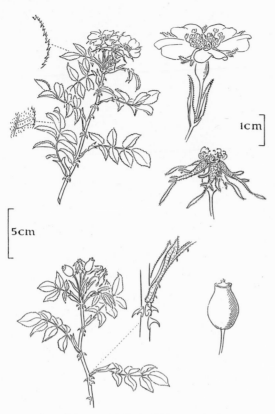

617.   *Rosa agrestis* Savi   White or pale pink

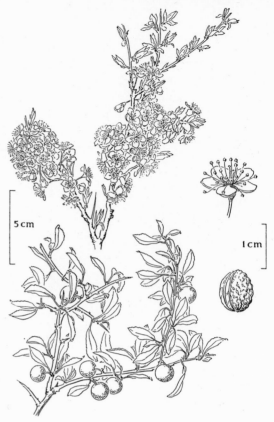

618.   *Prunus spinosa* L.   Blackthorn, Sloe   White

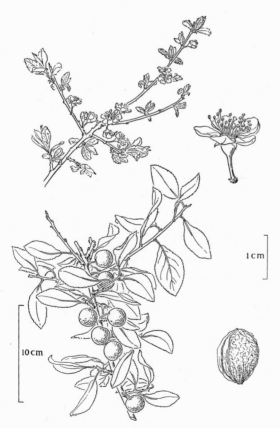

619.   *Prunus domestica* L. ssp. *insititia* (L.) C. K. Schneid.
Bullace   White

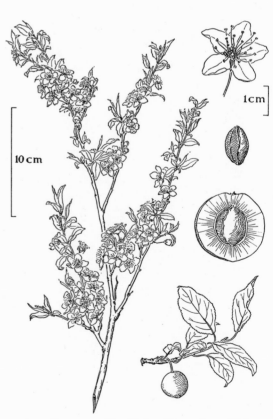

620.   *Prunus cerasifera* Ehrh.   Cherry-plum   White

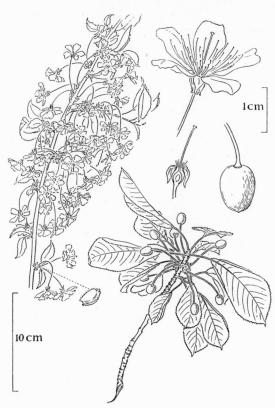

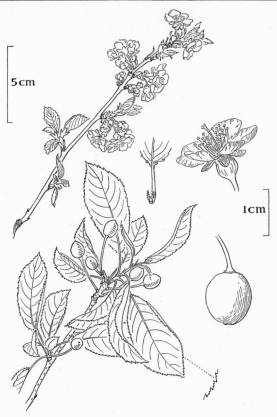

621. *Prunus avium* (L.) L.   Gean, Wild Cherry   White

622. *Prunus cerasus* L.   Sour Cherry   White

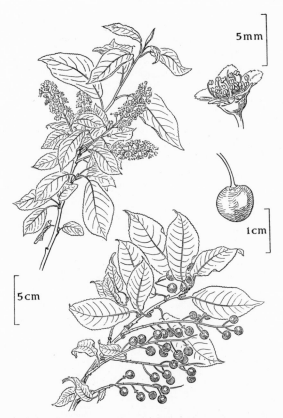

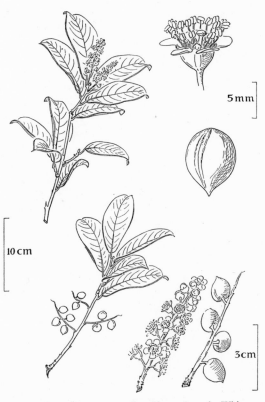

623. *Prunus padus* L.   Bird-Cherry   White

624. *Prunus laurocerasus* L.   Cherry-Laurel   White

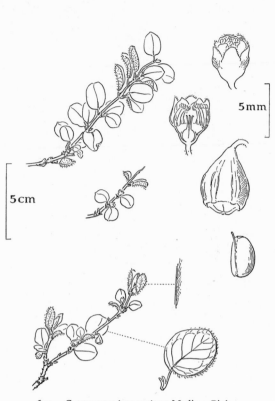

625. *Cotoneaster integerrimus* Medic.   Pink

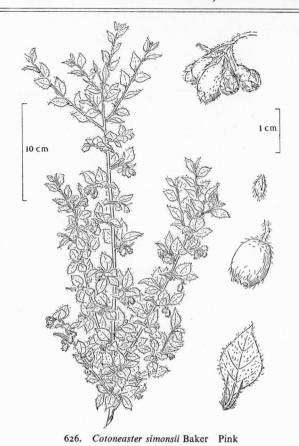

626. *Cotoneaster simonsii* Baker   Pink

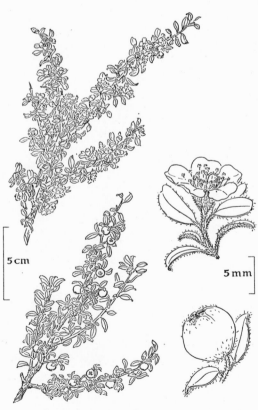

627. *Cotoneaster microphyllus* Wall. ex Lindl.   White

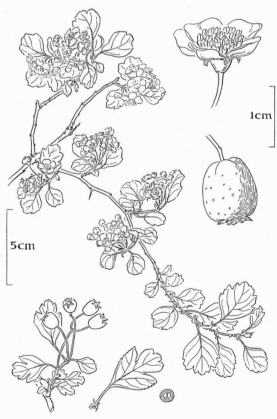

628. *Crataegus oxyacanthoides* Thuill.   Midland Hawthorn
White

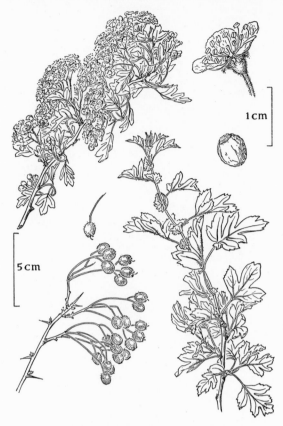

629. *Crataegus monogyna* Jacq. Hawthorn White

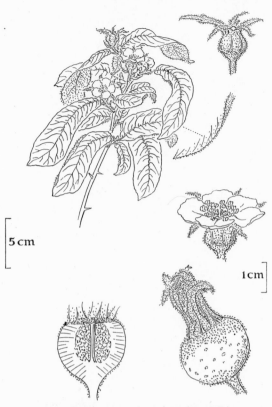

630. *Mespilus germanica* L. Medlar White

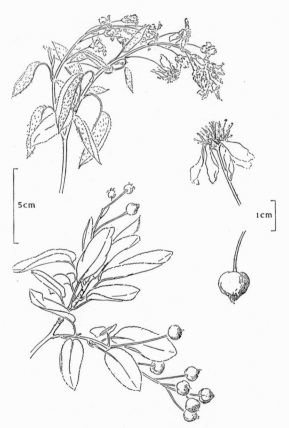

631. *Amelanchier laevis* Wieg. White

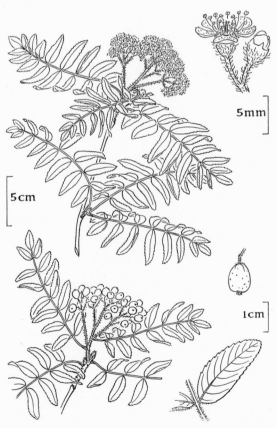

632 *Sorbus aucuparia* L. Rowan, Mountain Ash White

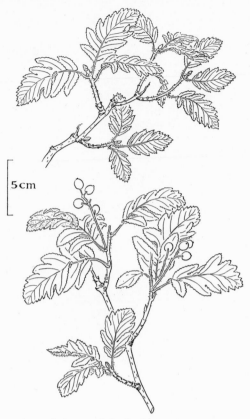

633.   *Sorbus pseudofennica* E. F. Warb.   White

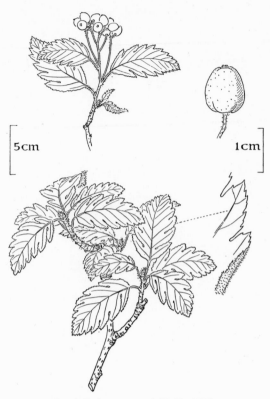

634.   *Sorbus arranensis* Hedl.   White

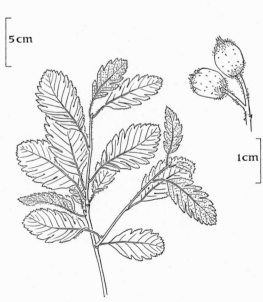

635.   *Sorbus leyana* Wilmott   White

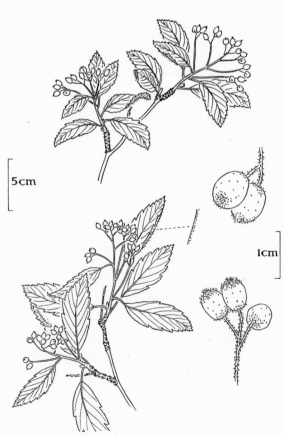

636.   *Sorbus minima* (A. Ley) Hedl.   White

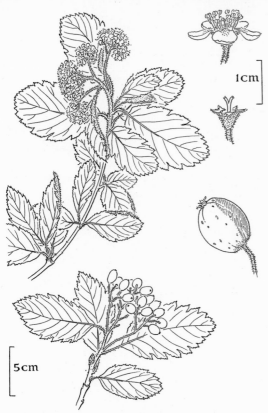

637. *Sorbus intermedia* (Ehrh.) Pers.   White

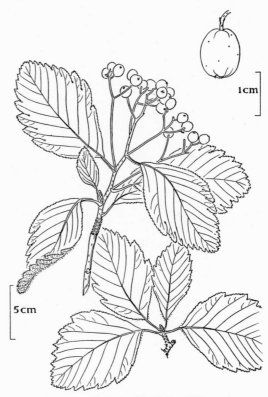

638. *Sorbus anglica* Hedl.   White

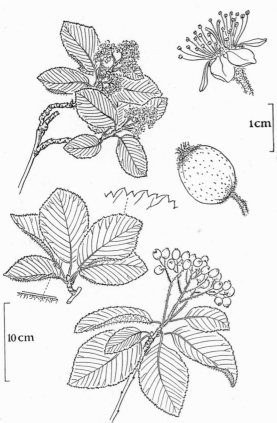

639. *Sorbus aria* (L.) Crantz   White Beam   White

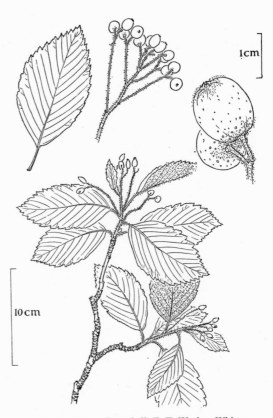

640. *Sorbus leptophylla* E. F. Warb.   White

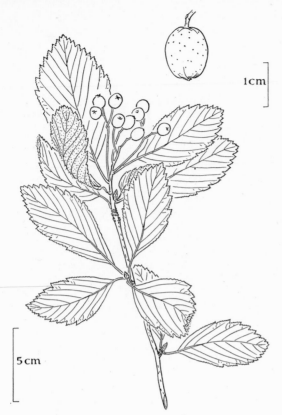

1cm

5cm

641. *Sorbus wilmottiana* E. F. Warb. White

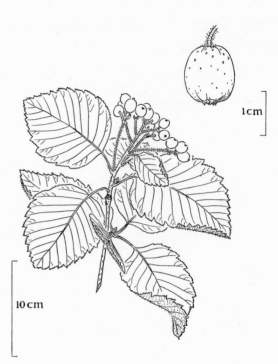

1cm

10cm

642. *Sorbus eminens* E. F. Warb. White

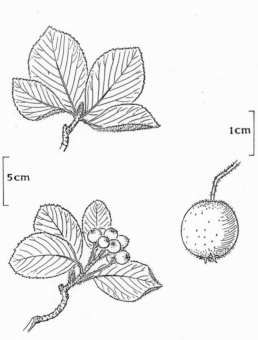

1cm

5cm

643. *Sorbus hibernica* E. F. Warb. White

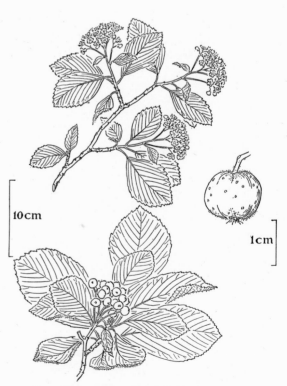

10cm

1cm

644. *Sorbus porrigentiformis* E. F. Warb. White

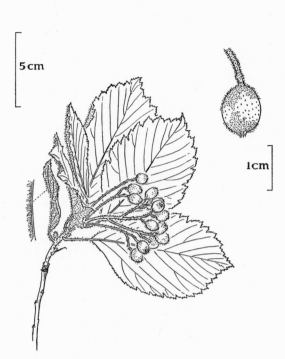

645. *Sorbus lancastriensis* E. F. Warb.   White

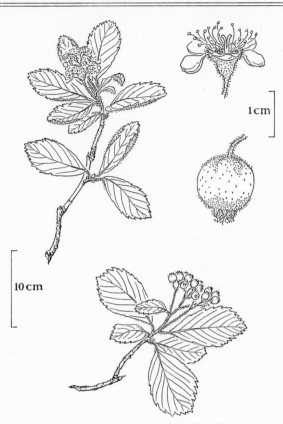

646. *Sorbus rupicola* (Syme) Hedl.   White

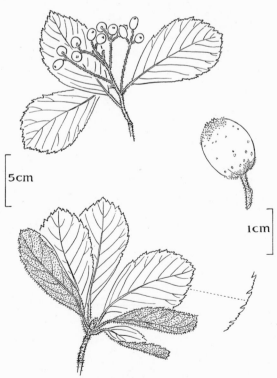

647. *Sorbus vexans* E. F. Warb.   White

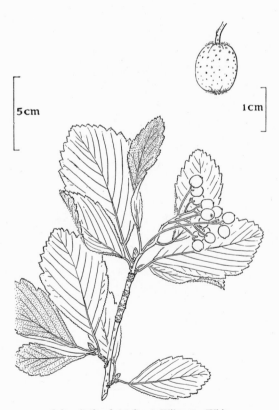

648. *Sorbus bristoliensis* Wilmott   White

649.   *Sorbus subcuneata* Wilmott   White

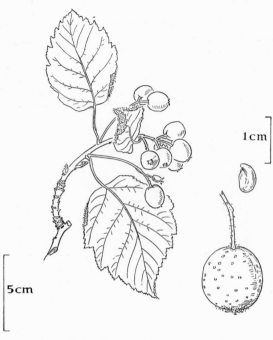

650.   *Sorbus devoniensis* E. F. Warb.   White

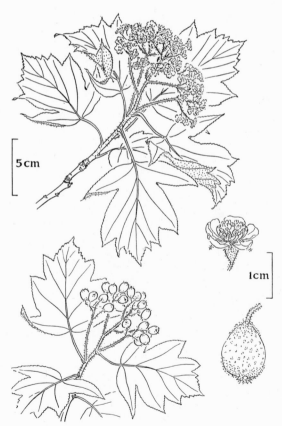

651.   *Sorbus torminalis* (L.) Crantz   Wild Service Tree
White

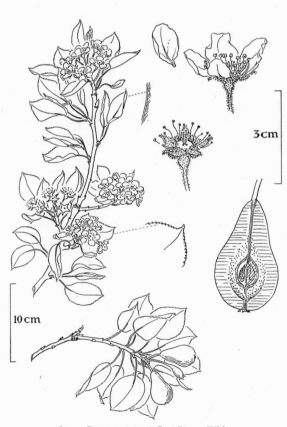

652.   *Pyrus communis* L.   Pear   White

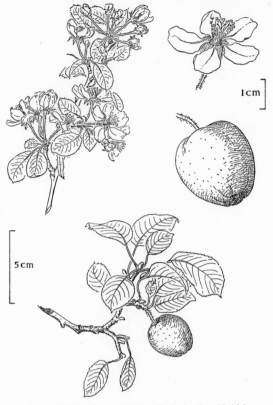

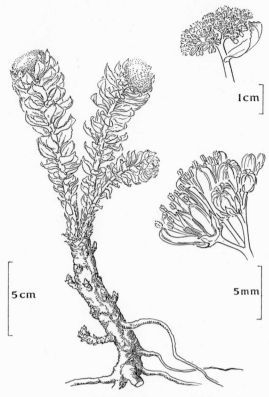

653. *Malus sylvestris* Mill. Crab Apple Pinkish

654. *Sedum rosea* (L.) Scop. Rose-root, Midsummer-men
Greenish

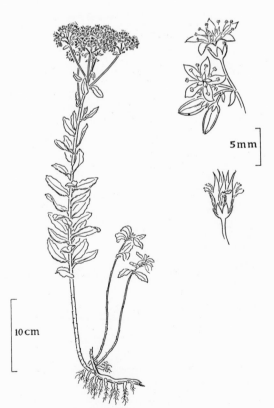

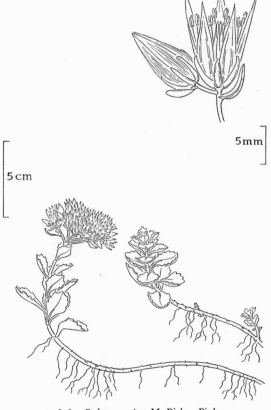

655. *Sedum telephium* L. Orpine, Livelong
Reddish-purple

656. *Sedum spurium* M. Bieb. Pink

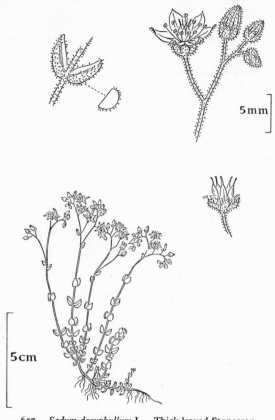

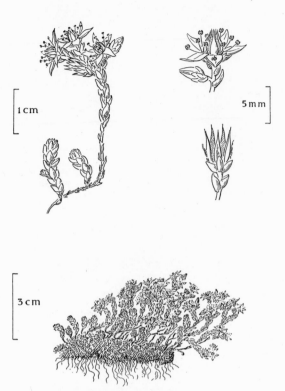

657. *Sedum dasyphyllum* L.   Thick-leaved Stonecrop
White

658. *Sedum anglicum* Huds.   English Stonecrop   White

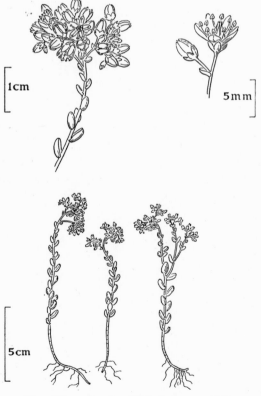

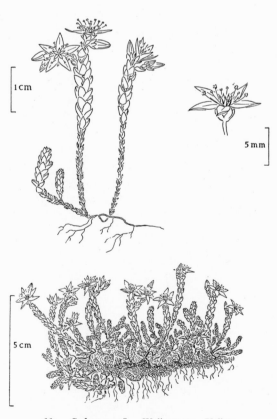

659. *Sedum album* L.   White Stonecrop   White

660. *Sedum acre* L.   Wall-pepper   Yellow

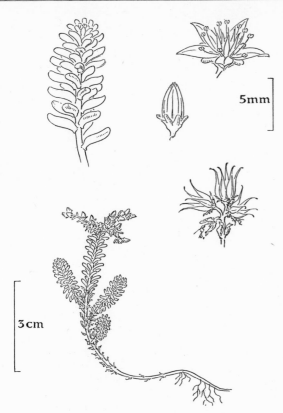

661. *Sedum sexangulare* L. Insipid Stonecrop Yellow

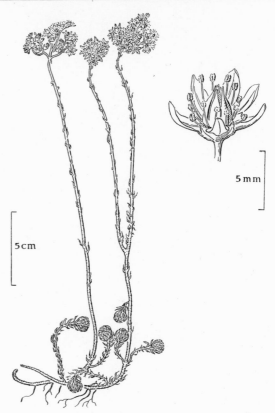

662. *Sedum forsteranum* Sm. Rock Stonecrop Yellow

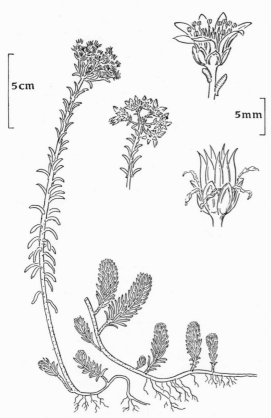

663. *Sedum reflexum* L. Yellow

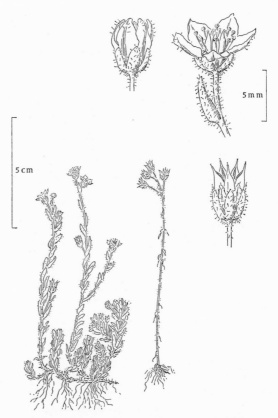

664. *Sedum villosum* L. Hairy Stonecrop Pink

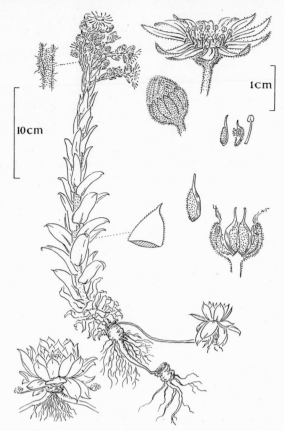

665. *Sempervivum tectorum* L. Houseleek Reddish

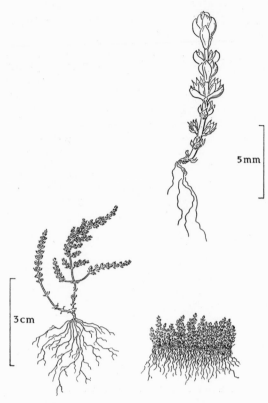

666. *Crassula tillaea* L.-Garland Whitish

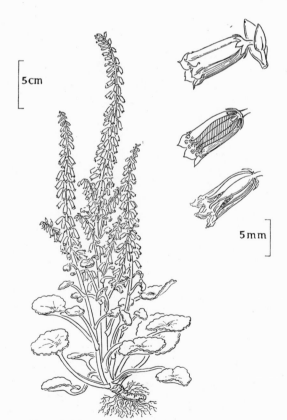

667. *Umbilicus rupestris* (Salisb.) Dandy Navelwort
Greenish-white

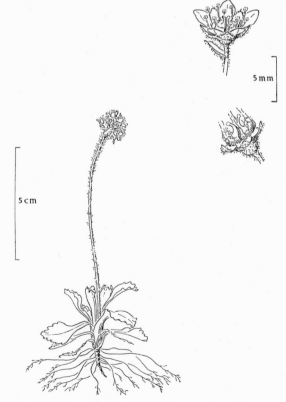

668. *Saxifraga nivalis* L. Alpine Saxifrage Greenish-white

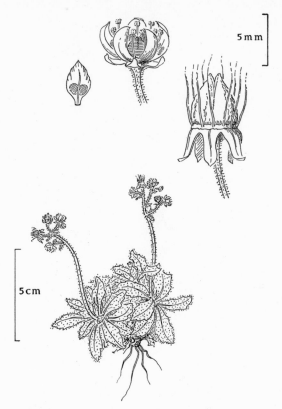

669. *Saxifraga stellaris* L.   Starry Saxifrage   White

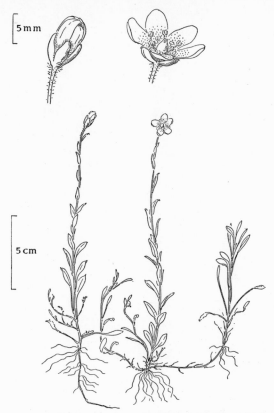

670. *Saxifraga hirculus* L.   Yellow Marsh Saxifrage   Yellow

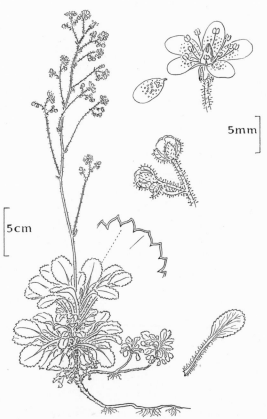

671. *Saxifraga spathularis × umbrosa*   London Pride   White

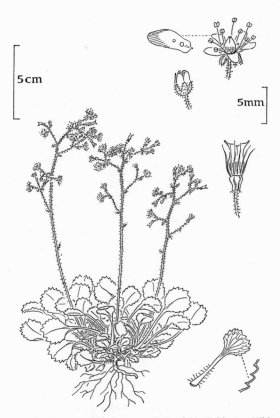

672. *Saxifraga spathularis* Brot.   St Patrick's Cabbage   White

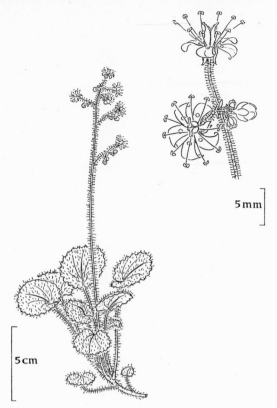

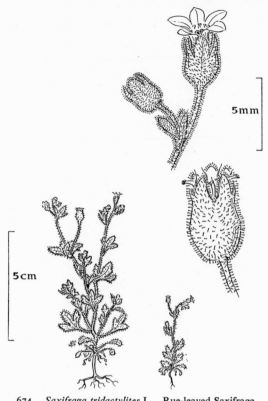

673. *Saxifraga hirsuta* L.   Kidney Saxifrage   White

674. *Saxifraga tridactylites* L.   Rue-leaved Saxifrage
White

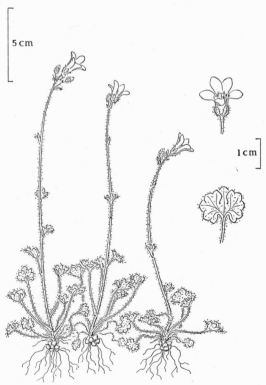

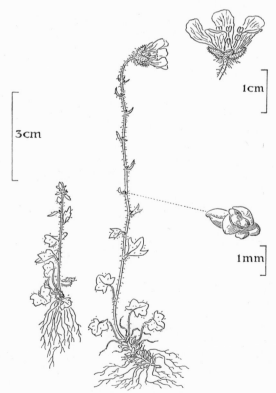

675. *Saxifraga granulata* L.   Meadow Saxifrage   White

676. *Saxifraga cernua* L.   Drooping Saxifrage   White

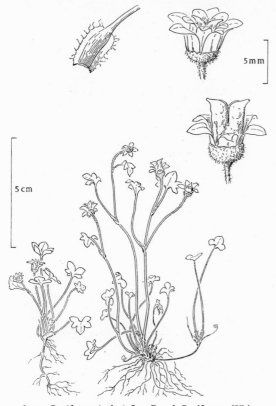

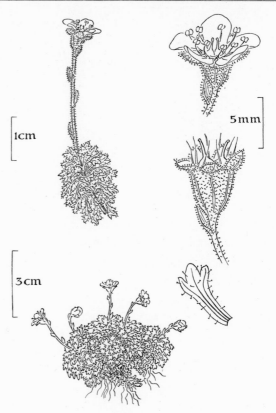

677. *Saxifraga rivularis* L. Brook Saxifrage White

678. *Saxifraga cespitosa* L. Tufted Saxifrage White

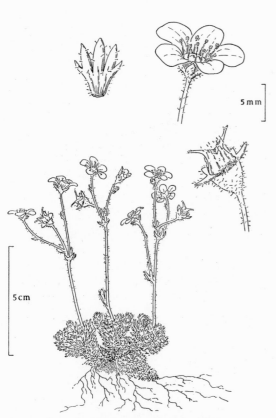

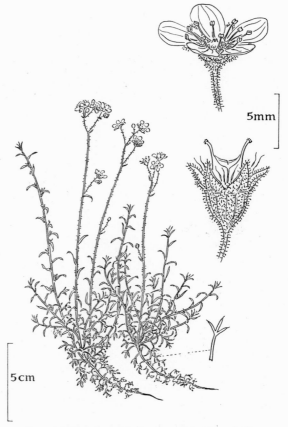

679. *Saxifraga rosacea* Moench White

680. *Saxifraga hypnoides* L. Dovedale Moss White

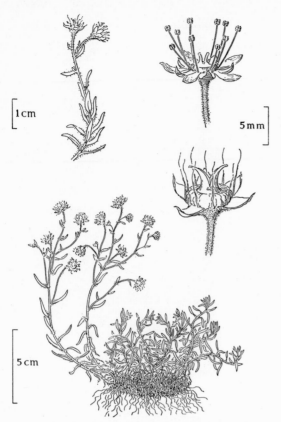

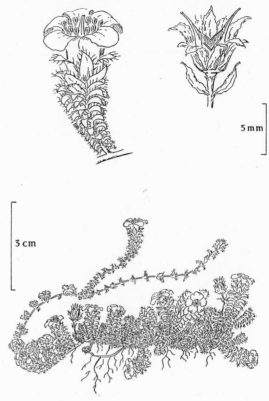

681.   *Saxifraga aizoides* L.   Yellow Mountain Saxifrage
Yellow

682.   *Saxifraga oppositifolia* L.   Purple Saxifrage   Purple

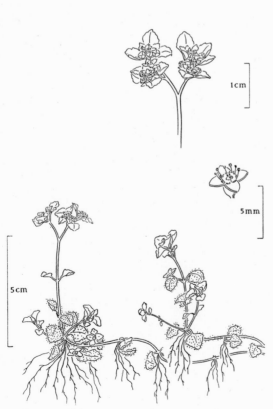

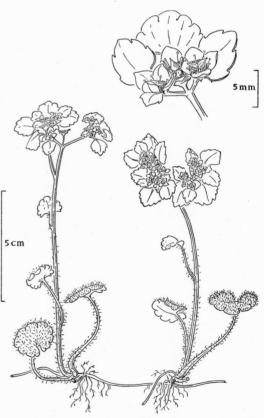

683.   *Chrysosplenium oppositifolium* L.   Opposite-leaved
Golden Saxifrage   Greenish-yellow

684.   *Chrysosplenium alternifolium* L.   Alternate-leaved
Golden Saxifrage   Greenish-yellow

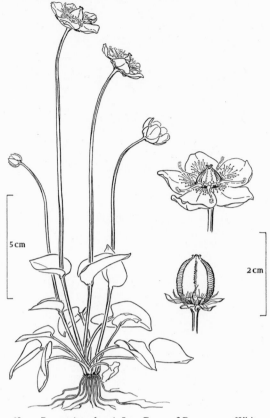

685.   *Parnassia palustris* L.   Grass of Parnassus   White

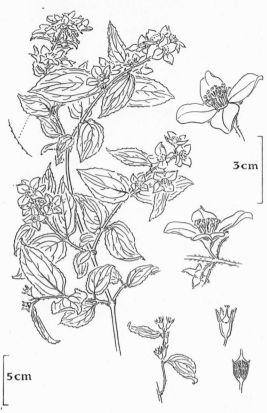

686.   *Philadelphus coronarius* L.   Syringa, Mock Orange
White

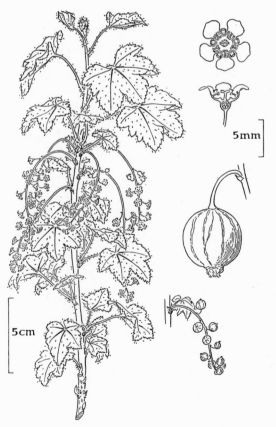

687.   *Ribes sylvestre* (Lam.) Mert. & Koch   Red Currant
Greenish

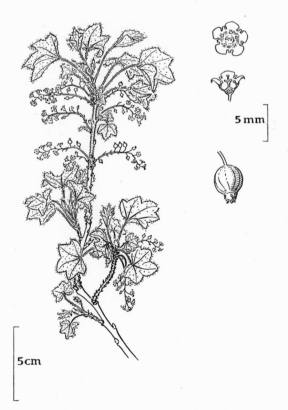

688.   *Ribes spicatum* Robson   Greenish

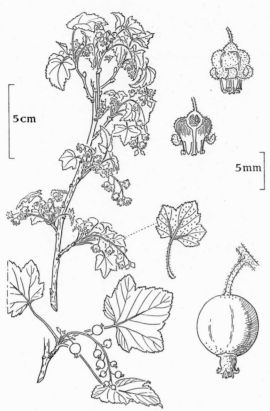

689. *Ribes nigrum* L.   Black Currant   Greenish

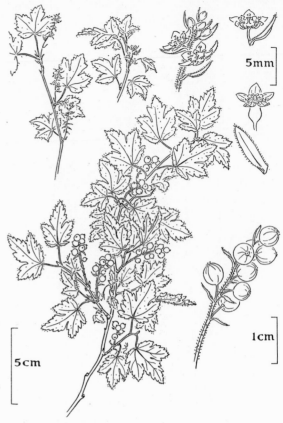

690. *Ribes alpinum* L.   Mountain Currant   Greenish

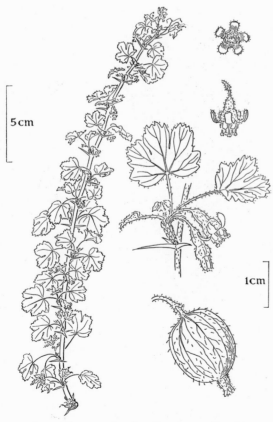

691. *Ribes uva-crispa* L.   Gooseberry   Greenish

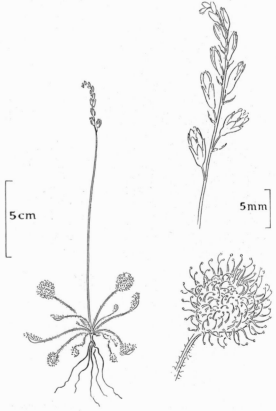

692. *Drosera rotundifolia* L.   Sundew   White

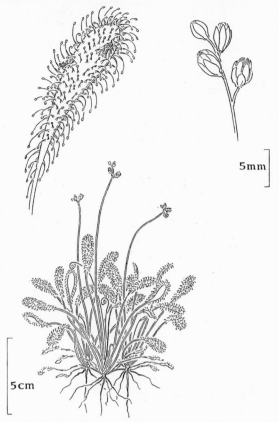

693.   *Drosera anglica* Huds.   Great Sundew   White

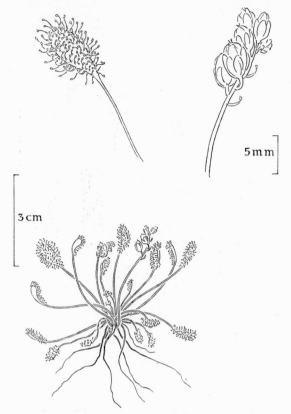

694.   *Drosera intermedia* Hayne   Long-leaved Sundew
White

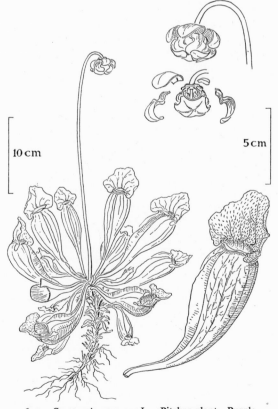

695.   *Sarracenia purpurea* L.   Pitcher-plant   Purple

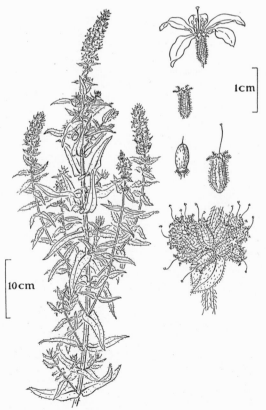

696.   *Lythrum salicaria* L.   Purple Loosestrife   Purple

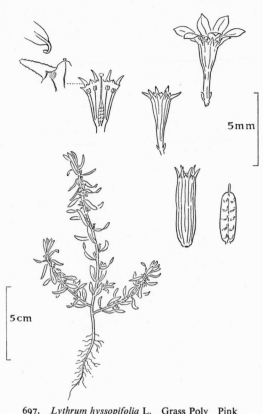

697. *Lythrum hyssopifolia* L. Grass Poly Pink

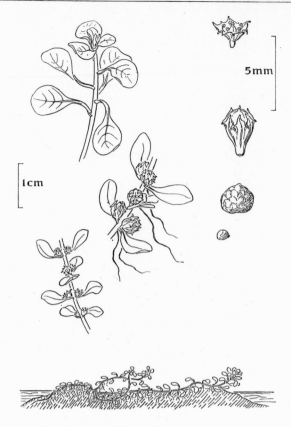

698. *Peplis portula* L. Water Purslane Greenish

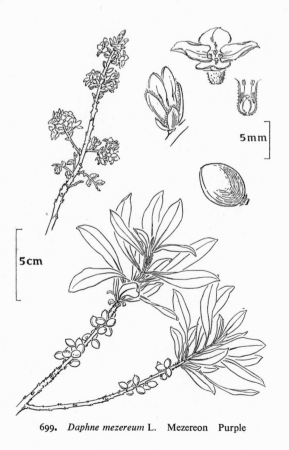

699. *Daphne mezereum* L. Mezereon Purple

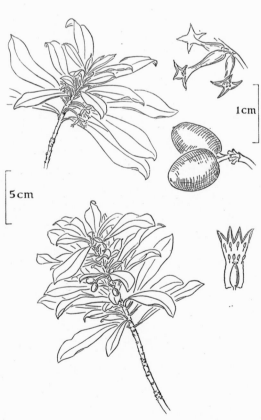

700. *Daphne laureola* L. Spurge-laurel Green

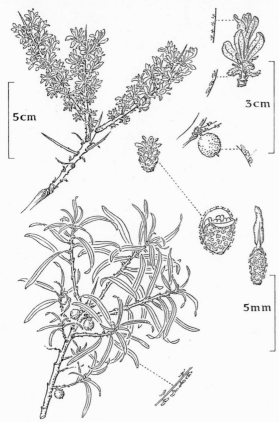

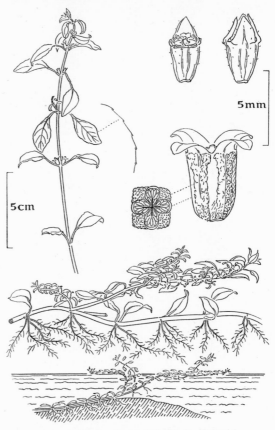

701. *Hippophaë rhamnoides* L.   Sea Buckthorn   Greenish

702. *Ludwigia palustris* (L.) Elliott   Greenish

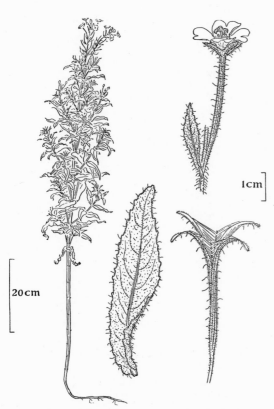

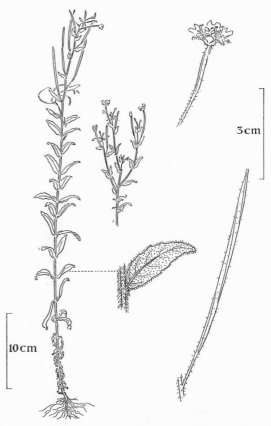

703. *Epilobium hirsutum* L.   Great Hairy Willow-herb, Codlins and Cream   Purplish

704. *Epilobium parviflorum* Schreb.   Small-flowered Hairy Willow-herb   Purplish

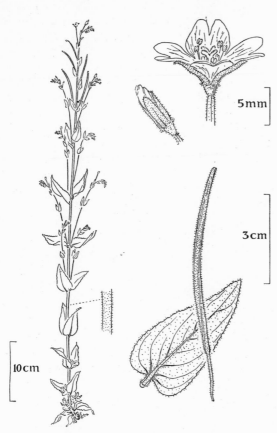

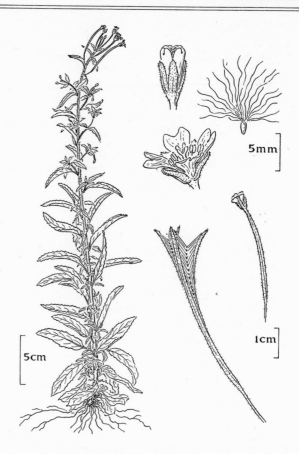

705. *Epilobium montanum* L.   Broad-leaved Willow-herb
Pink

706. *Epilobium lanceolatum* Seb. & Mauri   Spear-leaved
Willow-herb   Pink

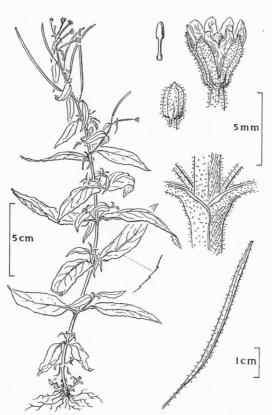

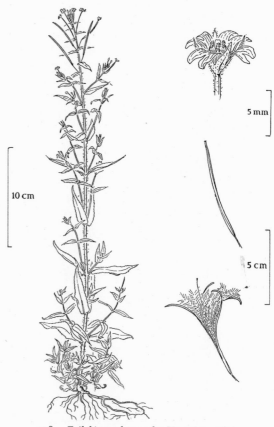

707. *Epilobium roseum* Schreb.   Small-flowered Willow-herb
White or pinkish

708. *Epilobium adenocaulon* Hausskn.   Pink

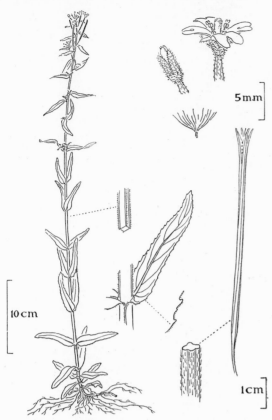

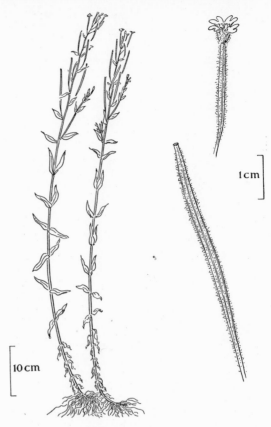

709.   *Epilobium adnatum* Griseb.   Square-stemmed Willow-
herb   Lilac

710.   *Epilobium obscurum* Schreb.   Pink

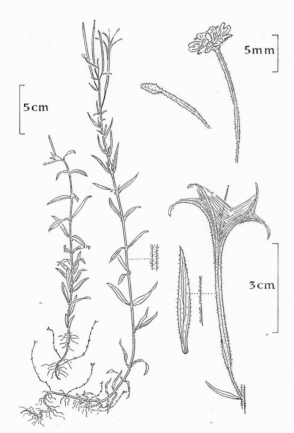

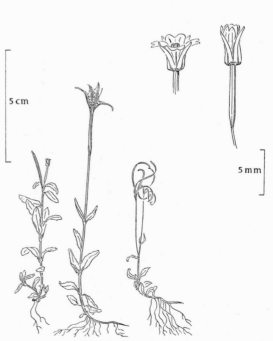

711.   *Epilobium palustre* L.   Marsh Willow-herb
Pink or lilac

712.   *Epilobium anagallidifolium* Lam.   Alpine Willow-herb
Pink

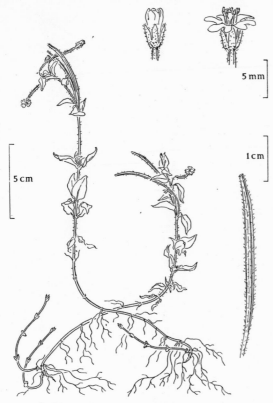

5 mm

5 cm

1 cm

713. *Epilobium alsinifolium* Vill. Chickweed Willow-herb
Reddish

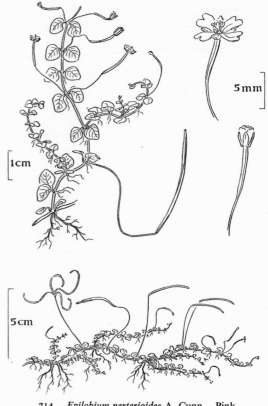

5 mm

1 cm

5 cm

714. *Epilobium nerterioides* A. Cunn. Pink

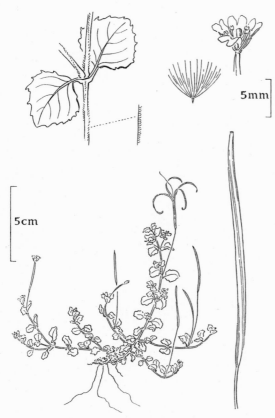

5 mm

5 cm

715. *Epilobium pedunculare* A. Cunn. Pink

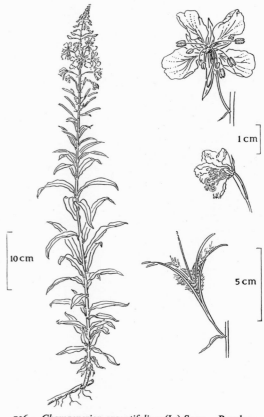

1 cm

10 cm

5 cm

716. *Chamaenerion angustifolium* (L.) Scop. Rosebay
Willow-herb, Fireweed Purple

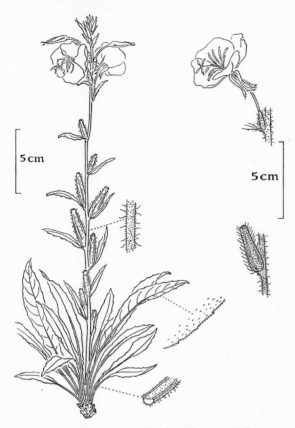

717.   *Oenothera biennis* L.   Evening Primrose   Yellow

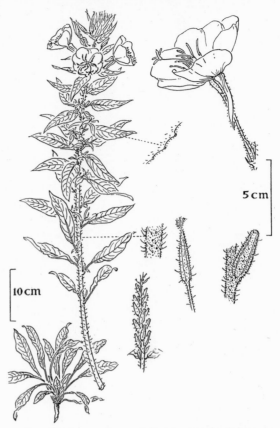

718.   *Oenothera erythrosepala* Borbás   Evening Primrose
Yellow

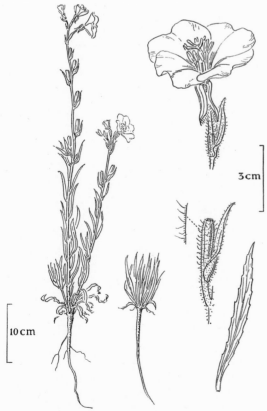

719.   *Oenothera stricta* Ledeb. ex Link   Evening Primrose
Yellow becoming red

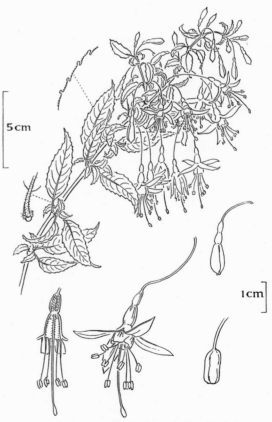

720.   *Fuchsia magellanica* Lam.   Willow Fuchsia
Red and violet

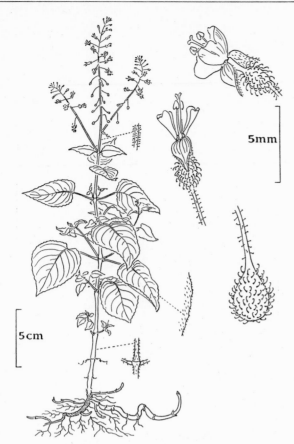

721. *Circaea lutetiana* L. Enchanter's Night-shade
White

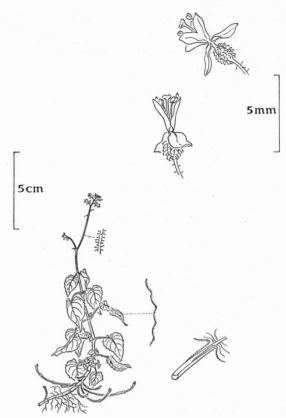

722. *Circaea alpina* L. Alpine Enchanter's Night-shade
White

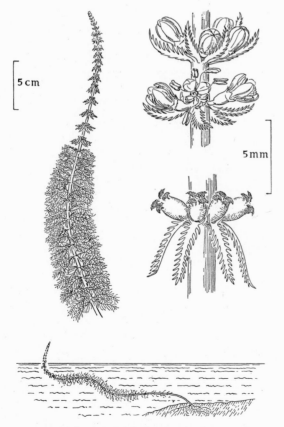

723. *Myriophyllum verticillatum* L. Whorled Water-milfoil
Greenish

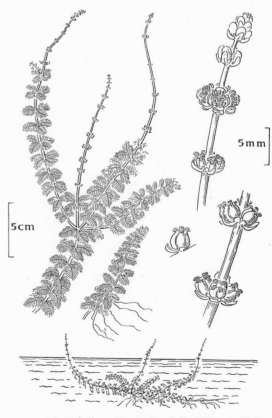

724. *Myriophyllum spicatum* L. Spiked Water-milfoil
Reddish

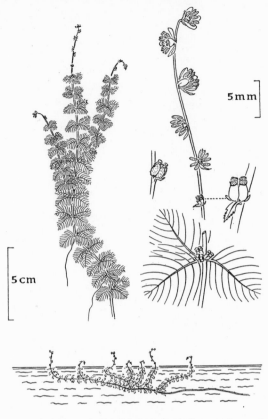

725. *Myriophyllum alterniflorum* DC. Alternate-flowered
Water-milfoil Yellowish

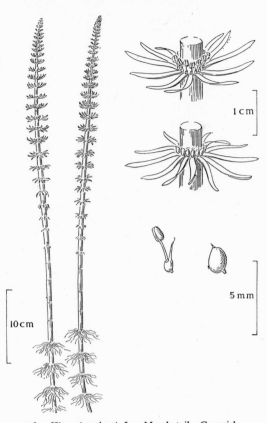

726. *Hippuris vulgaris* L. Mare's-tail Greenish

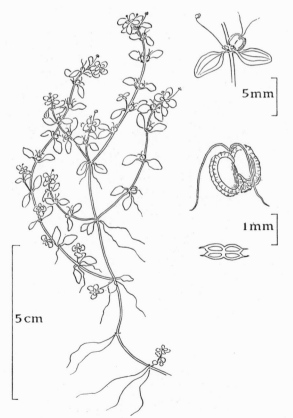

727. *Callitriche stagnalis* Scop. Greenish

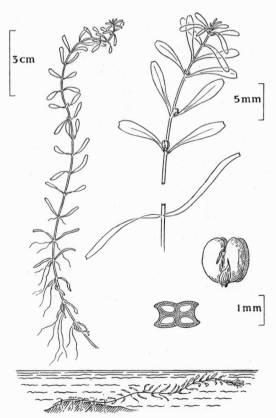

728. *Callitriche platycarpa* Kütz. Greenish

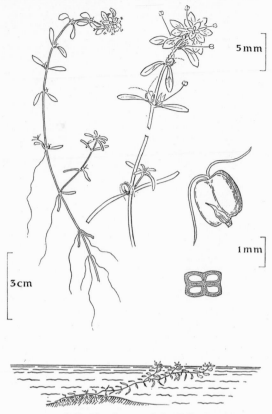

729. *Callitriche obtusangula* Le Gall   Greenish

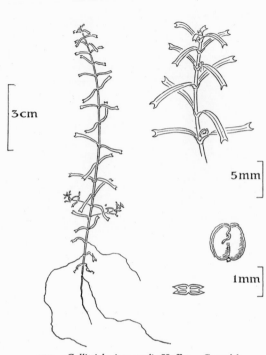

730. *Callitriche intermedia* Hoffm.   Greenish

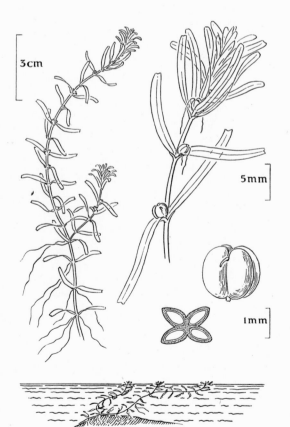

731. *Callitriche hermaphroditica* L.   Greenish

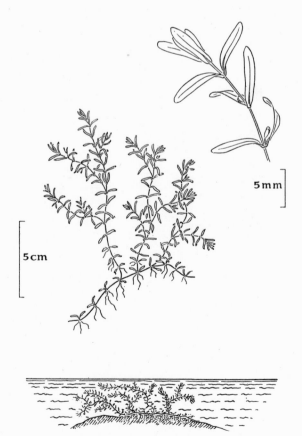

732. *Callitriche truncata* Guss.   Greenish

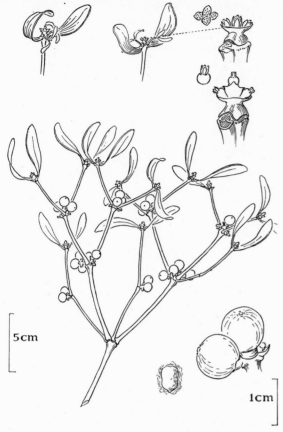

733.   *Viscum album* L.   Mistletoe   Greenish

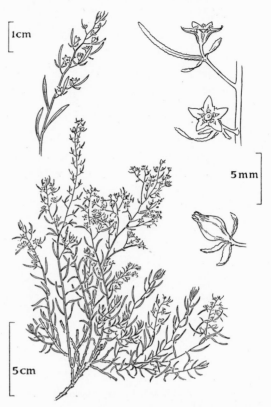

734.   *Thesium humifusum* DC.   Bastard Toadflax   Yellowish

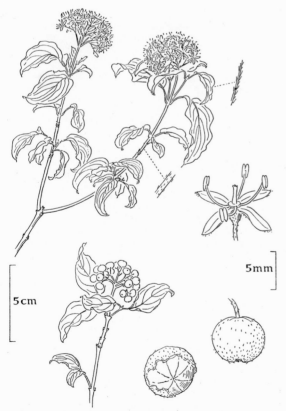

735.   *Thelycrania sanguinea* (L.) Fourr.   Dogwood   White

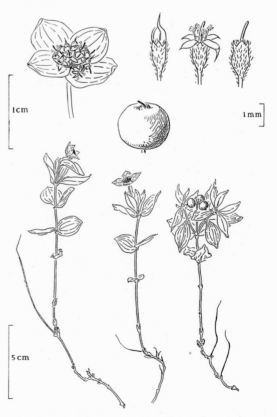

736.   *Chamaepericlymenum suecicum* (L.) Aschers. & Graebn.
Dwarf Cornel   White and dark purple

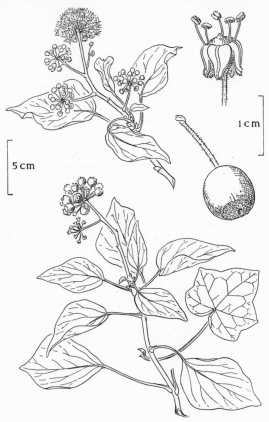

737. *Hedera helix* L. Ivy Yellowish

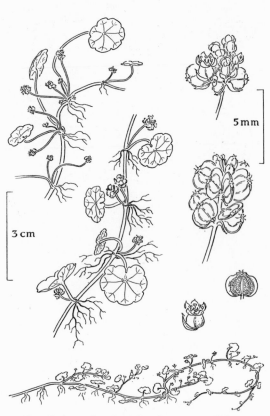

738. *Hydrocotyle vulgaris* L. Pennywort, White-rot
Greenish

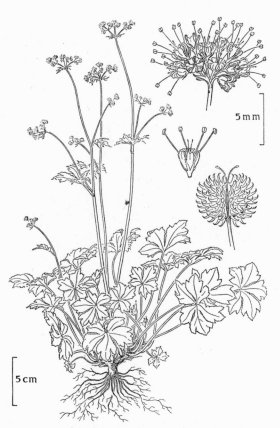

739. *Sanicula europaea* L. Sanicle Pink or white

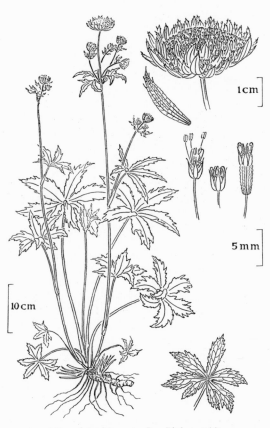

740. *Astrantia major* L. Pink or white

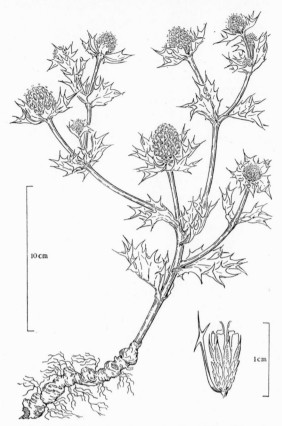

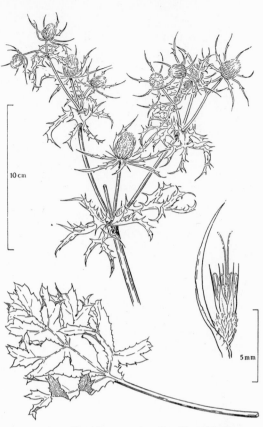

741. *Eryngium maritimum* L.   Sea Holly   Blue

742. *Eryngium campestre* L.   Greenish

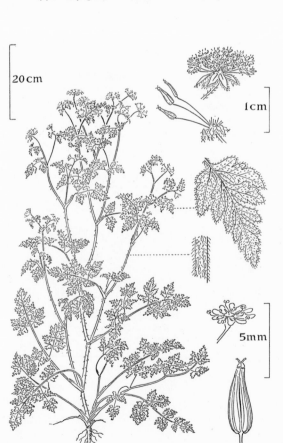

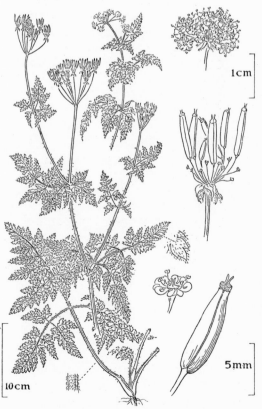

743. *Chaerophyllum temulentum* L.   Rough Chervil   White

744. *Chaerophyllum aureum* L.   White

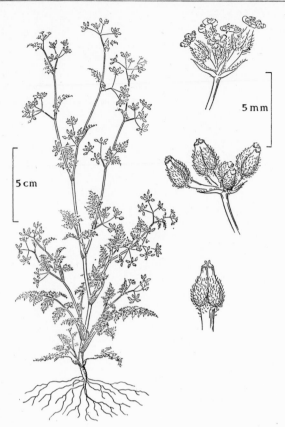

745.   *Anthriscus caucalis* M. Bieb.   Bur Chervil   White

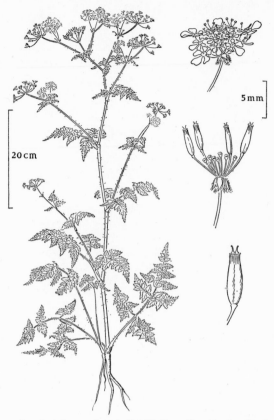

746.   *Anthriscus sylvestris* (L.) Hoffm.   Cow Parsley, Keck
White

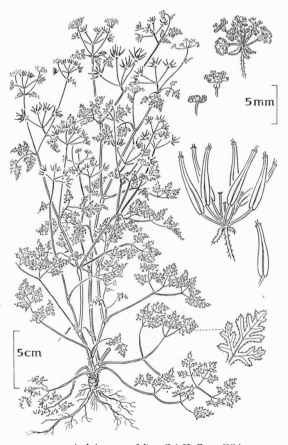

747.   *Anthriscus cerefolium* (L.) Hoffm.   White

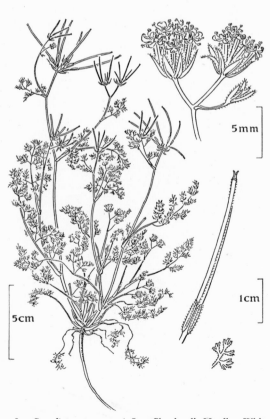

748.   *Scandix pecten-veneris* L.   Shepherd's Needle   White

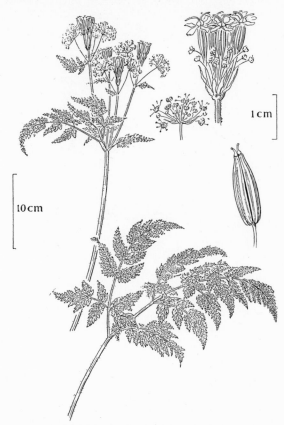

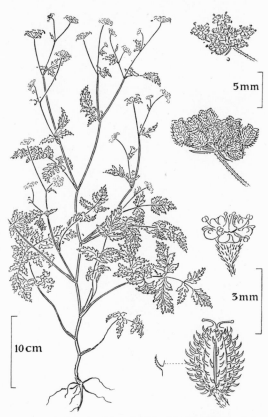

749. *Myrrhis odorata* (L.) Scop.   Sweet Cicely   White

750. *Torilis japonica* (Houtt.) DC.   Upright Hedge-parsley
White

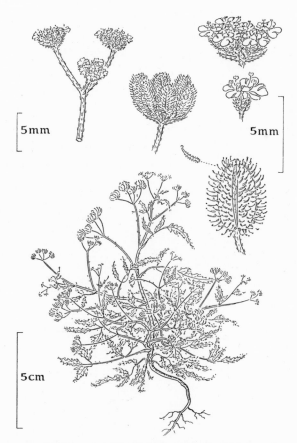

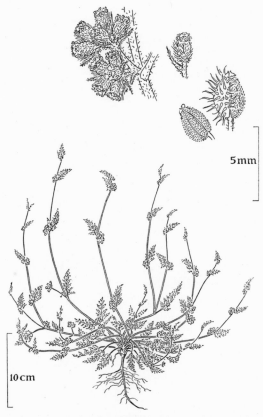

751. *Torilis arvensis* (Huds.) Link   Spreading Hedge-parsley
White

752. *Torilis nodosa* (L.) Gaertn.   Knotted Hedge-parsley
White

4-2

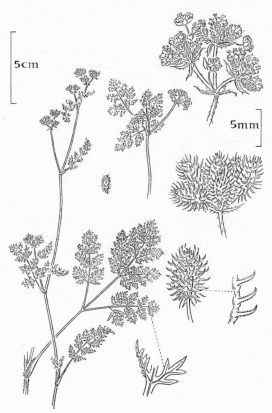

753. *Caucalis platycarpos* L. Small Bur-parsley White

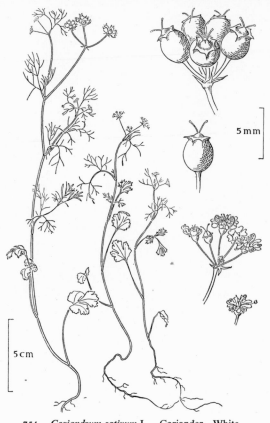

754. *Coriandrum sativum* L. Coriander White

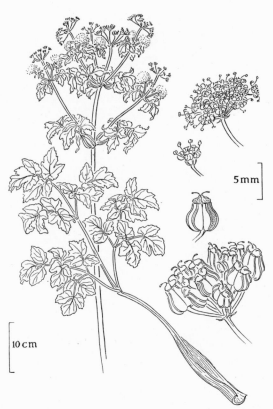

755. *Smyrnium olusatrum* L. Alexanders Yellowish

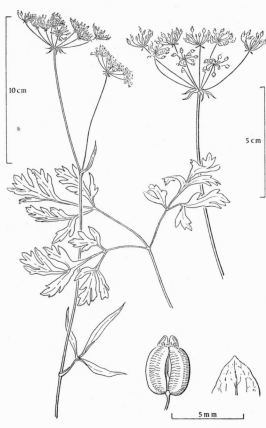

756. *Physospermum cornubiense* (L.) DC. Bladder-seed
White

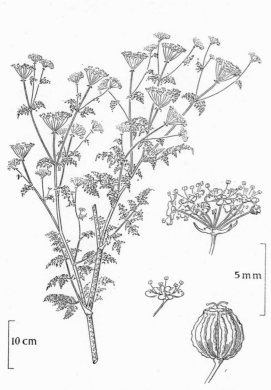

757.   *Conium maculatum* L.   Hemlock   White

758.   *Bupleurum fruticosum* L.   Yellowish

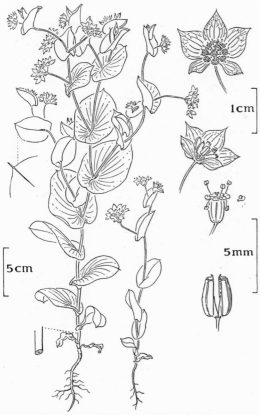

759.   *Bupleurum rotundifolium* L.   Hare's-ear, Thorow-wax
        Yellowish

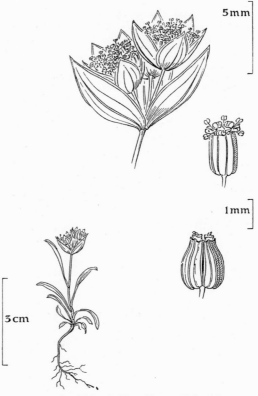

760.   *Bupleurum baldense* Turra   Yellowish

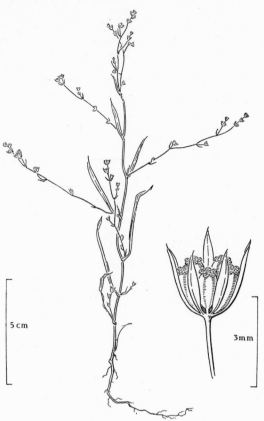

761.   *Bupleurum tenuissimum* L.   Smallest Hare's-ear
       Yellowish

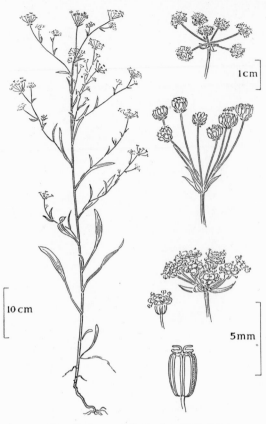

762.   *Bupleurum falcatum* L.   Yellowish

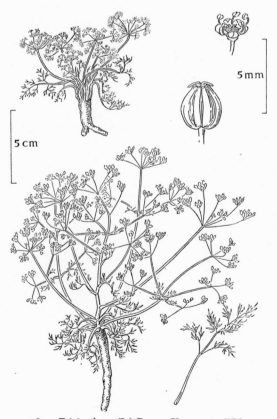

763.   *Trinia glauca* (L.) Dum.   Honewort   White

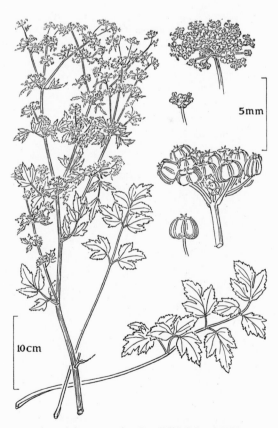

764.   *Apium graveolens* L.   Wild Celery   White

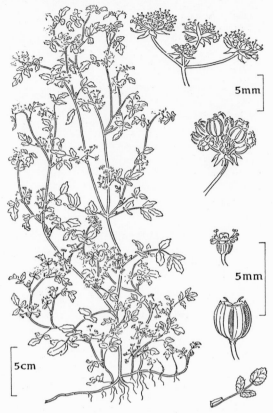

765.   *Apium nodiflorum* (L.) Lag.   Fool's Watercress   White

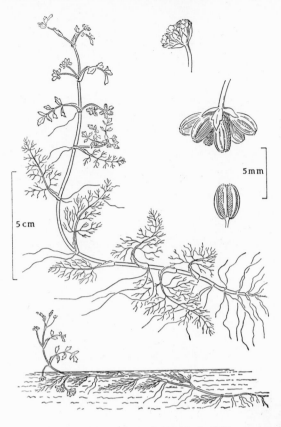

766.   *Apium inundatum* (L.) Rchb. f.   White

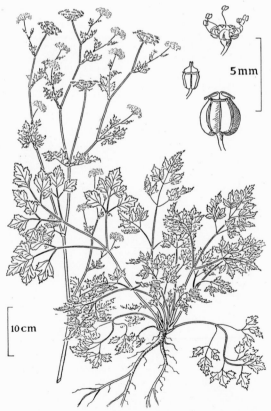

767.   *Petroselinum crispum* (Mill.) Nym.   Parsley   Yellowish

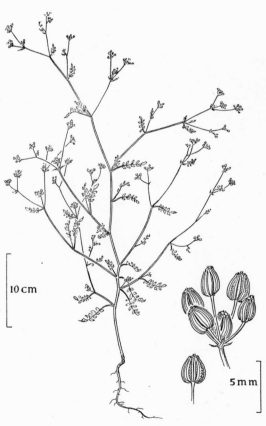

768.   *Petroselinum segetum* (L.) Koch   Corn Caraway
White

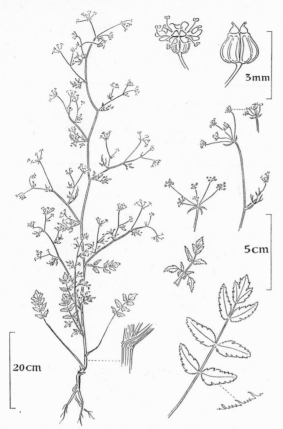

769.   *Sison amomum* L.   Stone Parsley   White

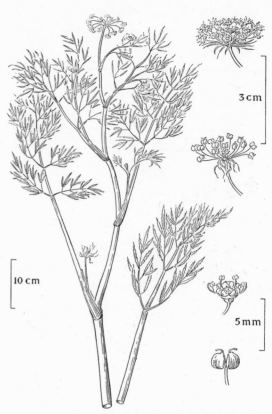

770.   *Cicuta virosa* L.   Cowbane   White

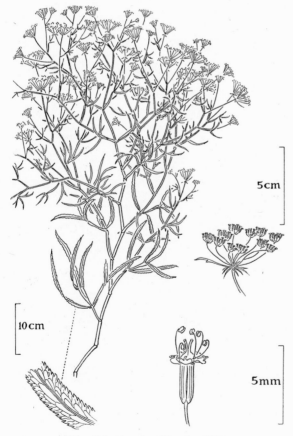

771.   *Falcaria vulgaris* Bernh.   White

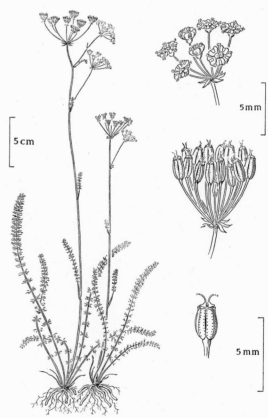

772.   *Carum verticillatum* (L.) Koch   Whorled Caraway
White

[56]

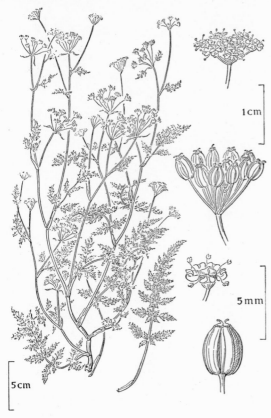

773.   *Carum carvi* L.   Caraway   White

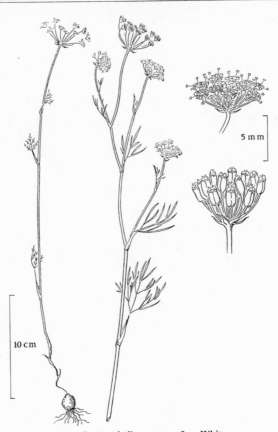

774.   *Bunium bulbocastanum* L.   White

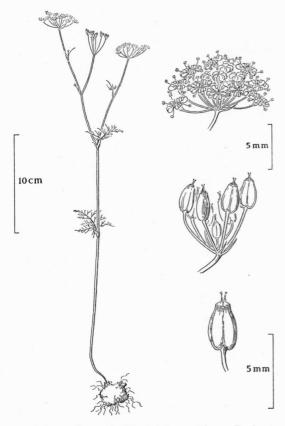

775.   *Conopodium majus* (Gouan) Loret   Pignut, Earthnut
White

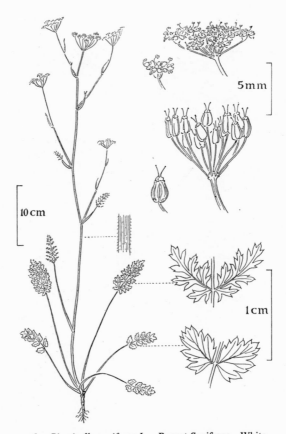

776.   *Pimpinella saxifraga* L.   Burnet Saxifrage   White

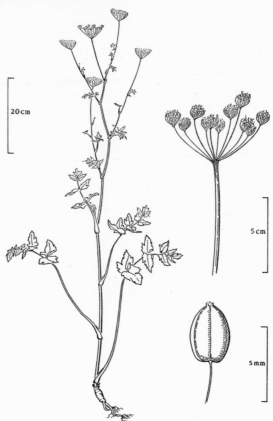

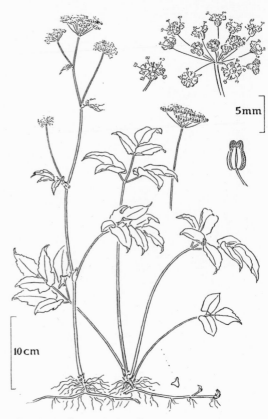

777.   *Pimpinella major* (L.) Huds.   Greater Burnet Saxifrage
White or pink

778.   *Aegopodium podagraria* L.   Goutweed, Ground Elder
White

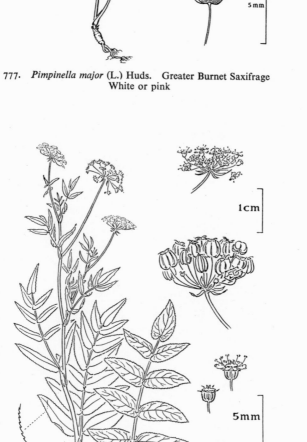

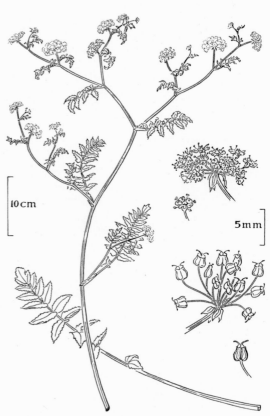

779.   *Sium latifolium* L.   Water Parsnip   White

780.   *Berula erecta* (Huds.) Coville   Narrow-leaved Water
Parsnip   White

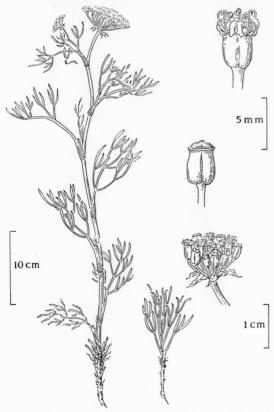

781. *Crithmum maritimum* L. Rock Samphire Yellowish

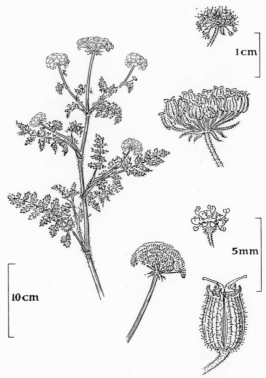

782. *Seseli libanotis* (L.) Koch White

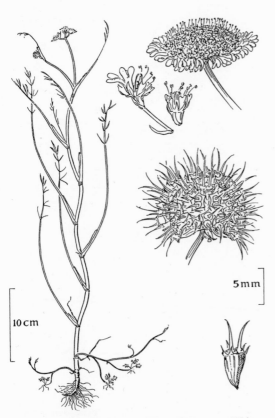

783. *Oenanthe fistulosa* L. Water Dropwort White

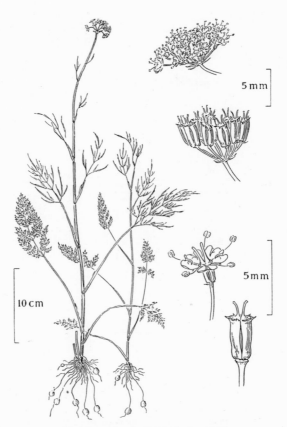

784. *Oenanthe pimpinelloides* L. White

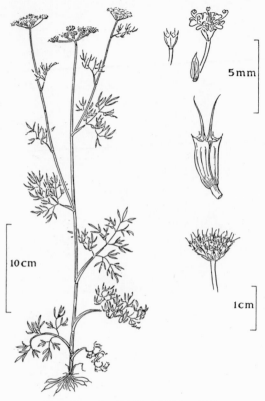

785.   *Oenanthe silaifolia* M. Bieb.   White

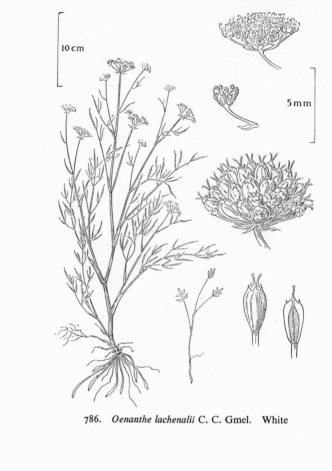

786.   *Oenanthe lachenalii* C. C. Gmel.   White

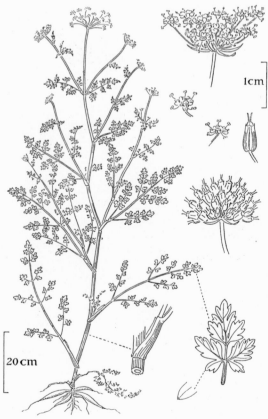

787.   *Oenanthe crocata* L.   Hemlock Water Dropwort
       White

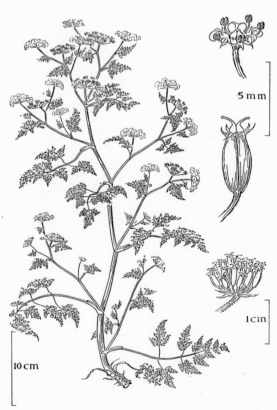

788.   *Oenanthe aquatica* (L.) Poir.   Fine-leaved Water
       Dropwort   White

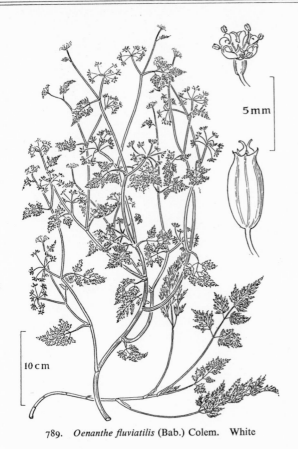

789. *Oenanthe fluviatilis* (Bab.) Colem. White

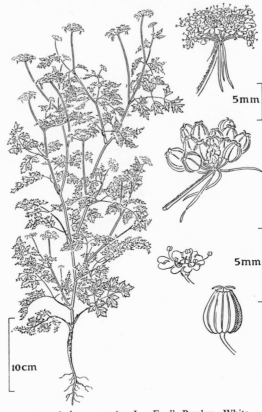

790. *Aethusa cynapium* L. Fool's Parsley White

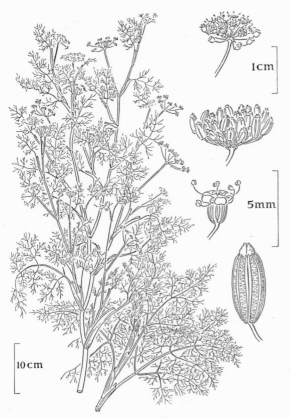

791. *Foeniculum vulgare* Mill. Fennel Yellow

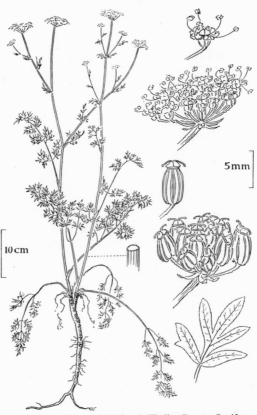

792. *Silaum silaus* (L.) Schinz & Thell. Pepper Saxifrage
Yellowish

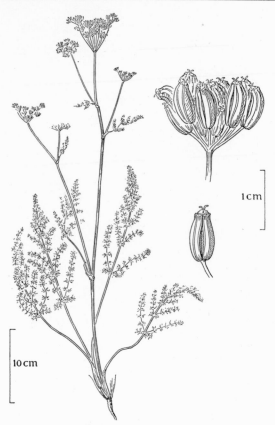

793. *Meum athamanticum* Jacq. Spignel, Meu, Baldmoney
White

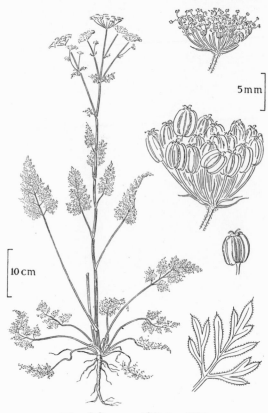

794. *Selinum carvifolia* L. White

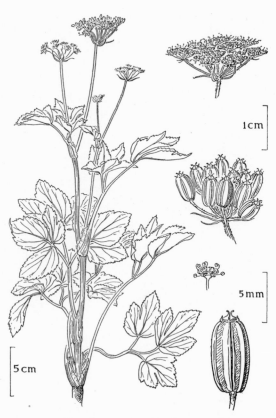

795. *Ligusticum scoticum* L. Lovage White

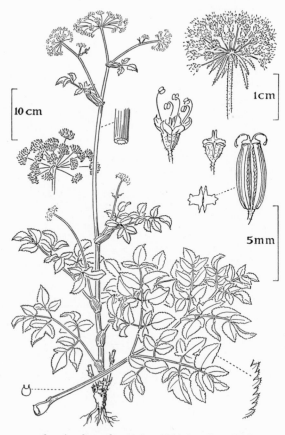

796. *Angelica sylvestris* L. Wild Angelica White

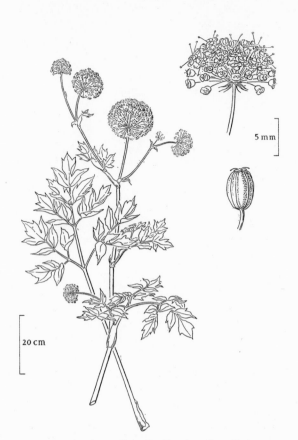

797. *Angelica archangelica* L.   Angelica   Green

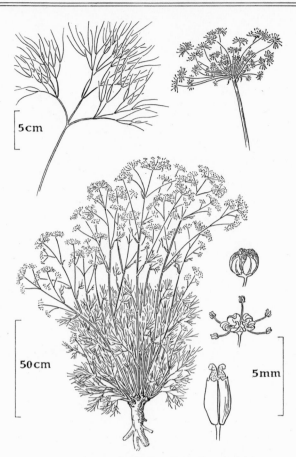

798. *Peucedanum officinale* L.   Hog's Fennel, Sulphur-
weed   Yellow

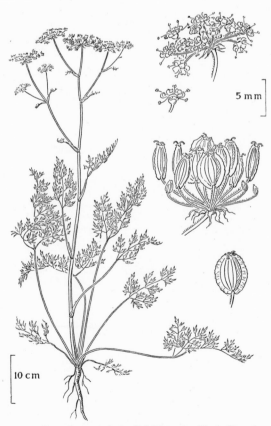

799. *Peucedanum palustre* (L.) Moench   Hog's Fennel,
Milk Parsley   White

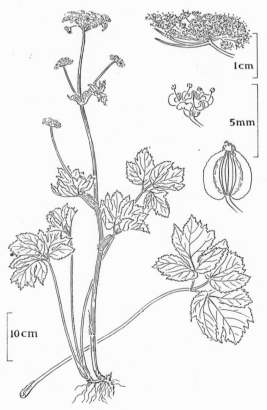

800. *Peucedanum ostruthium* (L.) Koch   Masterwort   White

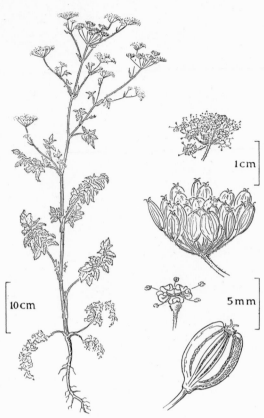

801.   *Pastinaca sativa* L.   Wild Parsnip   Yellow

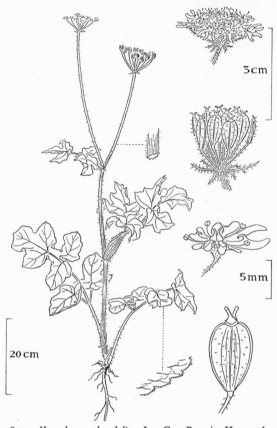

802.   *Heracleum sphondylium* L.   Cow Parsnip, Hogweed, Keck   Whitish

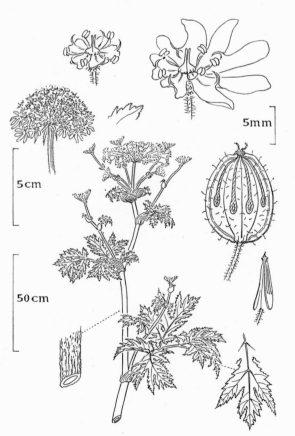

803.   *Heracleum mantegazzianum* Somm. & Lev.   White

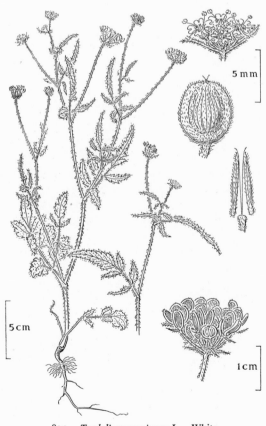

804.   *Tordylium maximum* L.   White

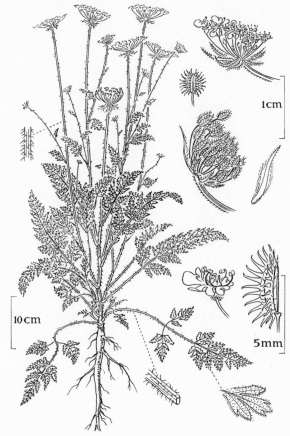

805. *Daucus carota* L. Wild carrot White

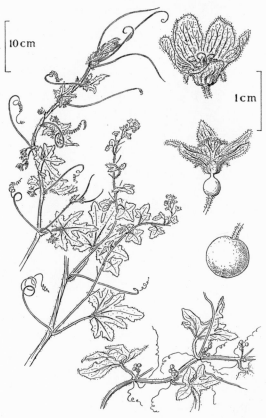

806. *Bryonia dioica* Jacq. White Bryony Greenish

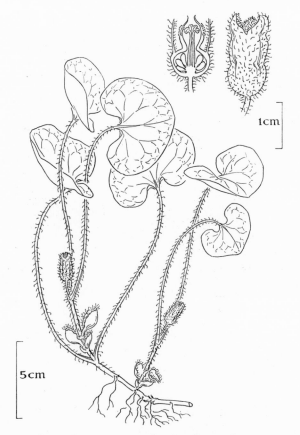

807. *Asarum europaeum* L. Asarabacca Brown

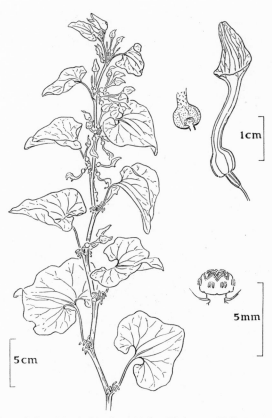

808. *Aristolochia clematitis* L. Birthwort Yellow

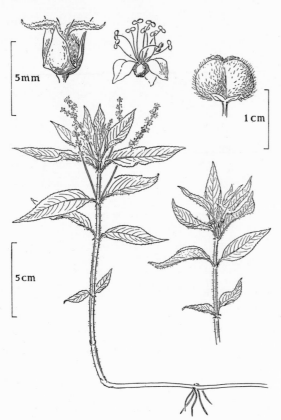

809. *Mercurialis perennis* L.   Dog's Mercury   Green

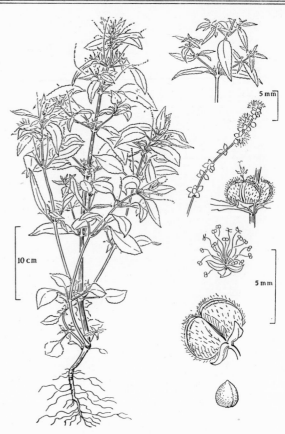

810. *Mercurialis annua* L.   Annual Mercury   Green

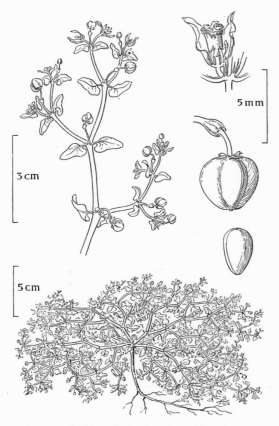

811. *Euphorbia peplis* L.   Purple Spurge   Greenish

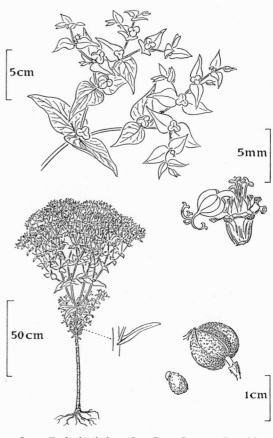

812. *Euphorbia lathyrus* L.   Caper Spurge   Greenish

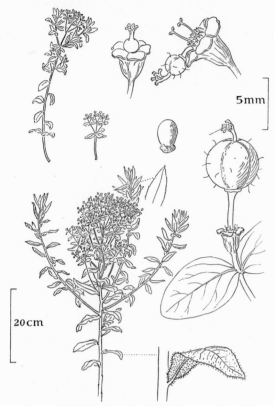

813. *Euphorbia pilosa* L.   Hairy Spurge   Greenish

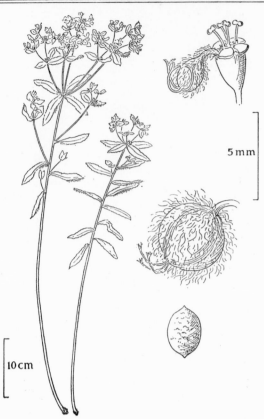

814. *Euphorbia corallioides* L.   Coral Spurge   Greenish

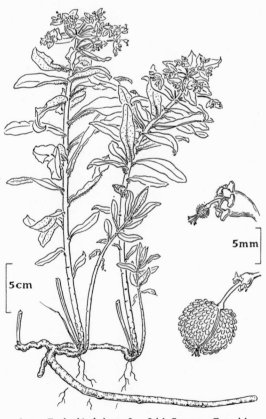

815. *Euphorbia hyberna* L.   Irish Spurge   Greenish

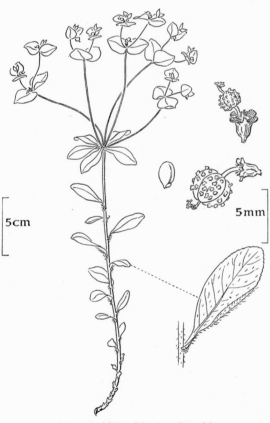

816. *Euphorbia dulcis* L.   Greenish

[67]

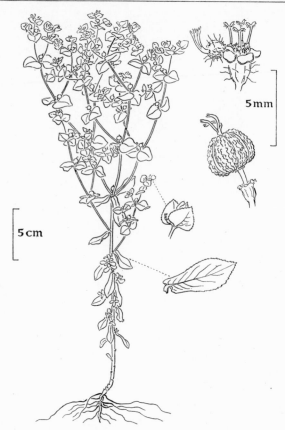

5mm

5cm

817. *Euphorbia platyphyllos* L. Broad Spurge Greenish

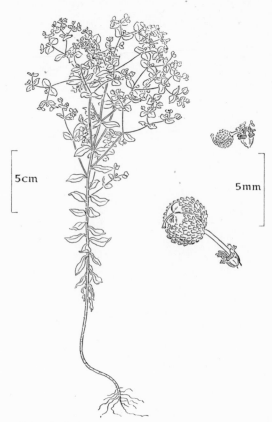

5cm

5mm

818. *Euphorbia stricta* L. Upright Spurge Greenish

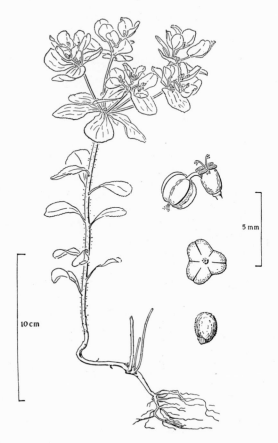

5mm

10cm

819. *Euphorbia helioscopia* L. Sun Spurge Greenish

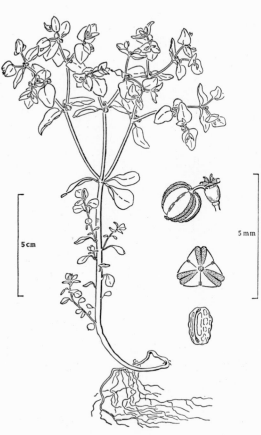

5cm

5mm

820. *Euphorbia peplus* L. Petty Spurge Greenish

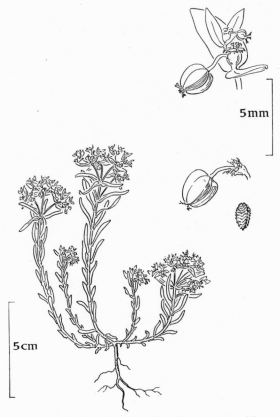

821. *Euphorbia exigua* L.   Dwarf Spurge   Greenish

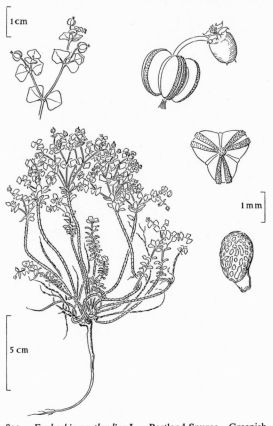

822. *Euphorbia portlandica* L.   Portland Spurge   Greenish

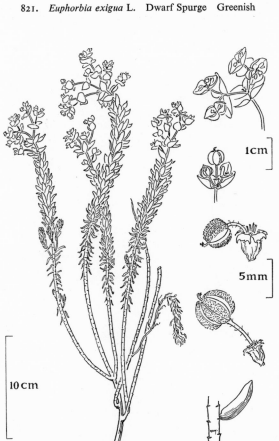

823. *Euphorbia paralias* L.   Sea Spurge   Greenish

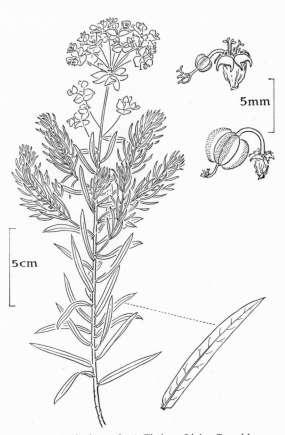

824. *Euphorbia uralensis* Fisch. ex Link   Greenish

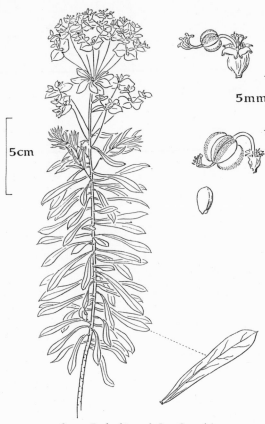

825. *Euphorbia esula* L.   Greenish

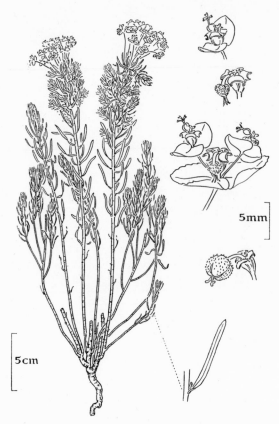

826. *Euphorbia cyparissias* L.   Cypress Spurge   Greenish

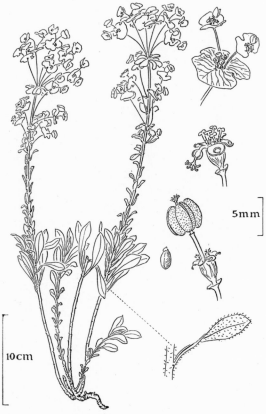

827. *Euphorbia amygdaloides* L.   Wood Spurge   Greenish

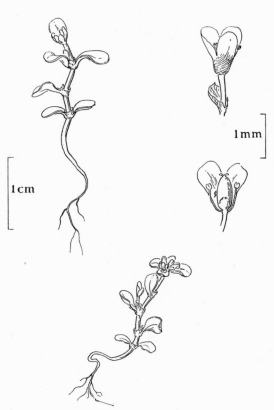

828. *Koenigia islandica* L.   Greenish

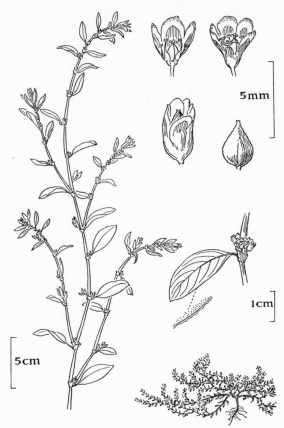

829. *Polygonum aviculare* L. Knotgrass Greenish

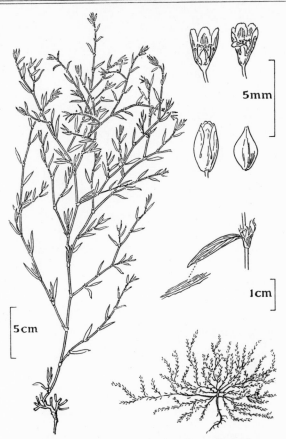

830. *Polygonum rurivagum* Jord. ex Bor. Greenish

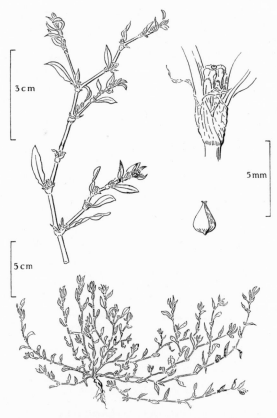

831. *Polygonum arenastrum* Bor. Knotgrass Greenish

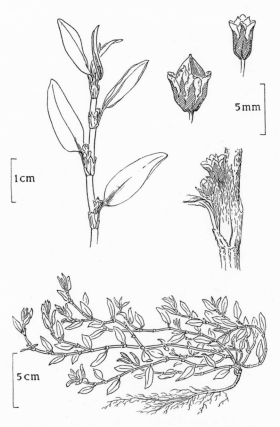

832. *Polygonum raii* Bab. Ray's Knotgrass Greenish

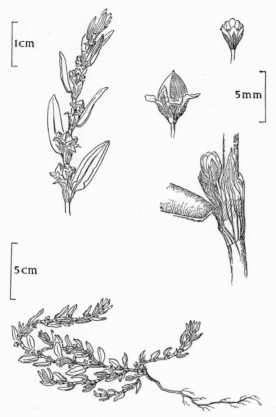

1 cm

5 mm

5 cm

5 cm

833. *Polygonum maritimum* L. Sea Knotgrass Pinkish

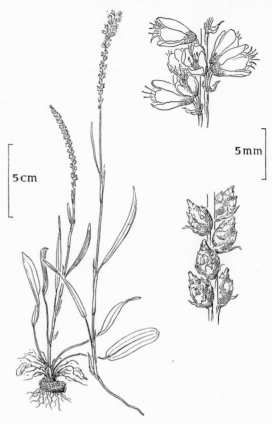

5 cm

5 mm

834. *Polygonum viviparum* L. White

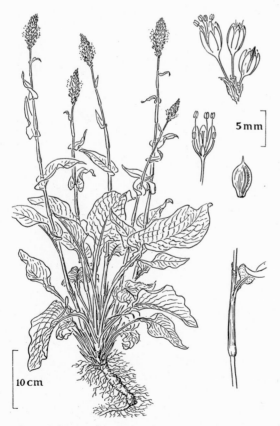

5 mm

10 cm

835. *Polygonum bistorta* L. Snake-root, Easter-ledges,
Bistort Pink

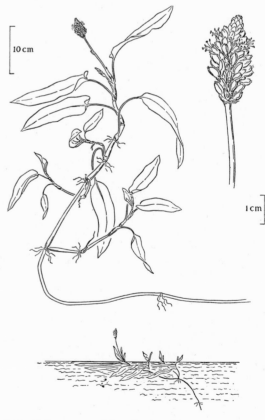

10 cm

1 cm

836. *Polygonum amphibium* L. Amphibious Bistort Pink
(Water form)

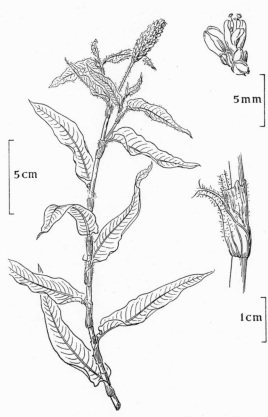

837. *Polygonum amphibium* L.   Amphibious Bistort   Pink
(Land form)

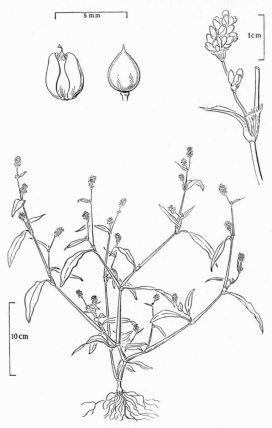

838. *Polygonum persicaria* L.   Persicaria   Pink

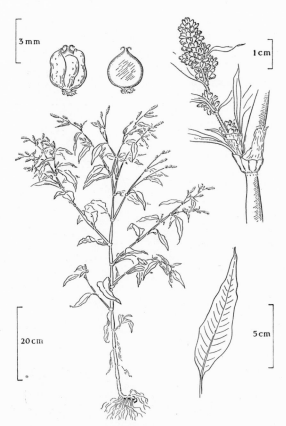

839. *Polygonum lapathifolium* L.   Pale Persicaria
Greenish-white

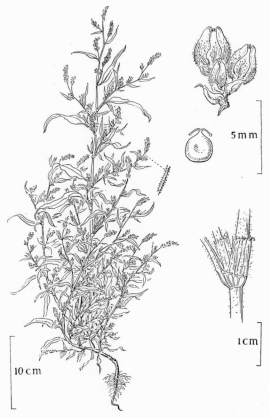

840. *Polygonum nodosum* Pers.   Pink

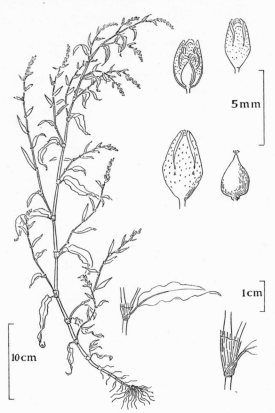

841.   *Polygonum hydropiper* L.   Water-pepper   Greenish

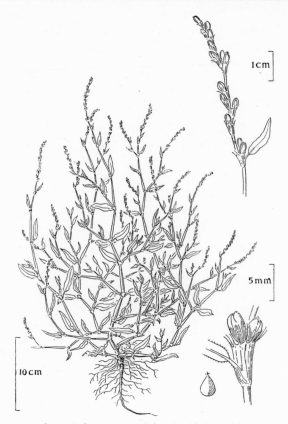

842.   *Polygonum mite* Schrank   Pink or white

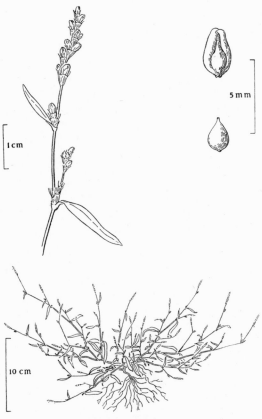

843.   *Polygonum minus* Huds.   Pink or white

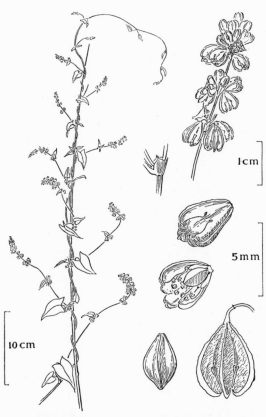

844.   *Polygonum convolvulus* L.   Black Bindweed   White

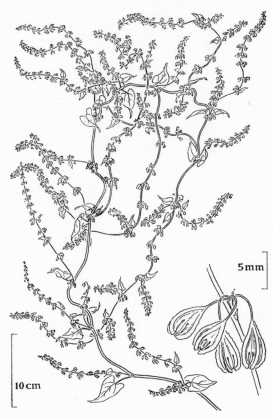

845.   *Polygonum dumetorum* L.   White

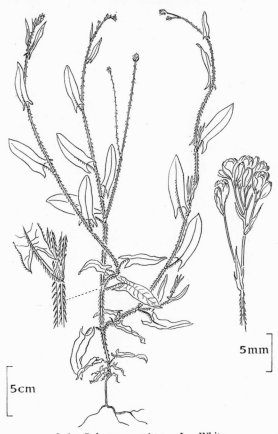

846.   *Polygonum sagittatum* L.   White

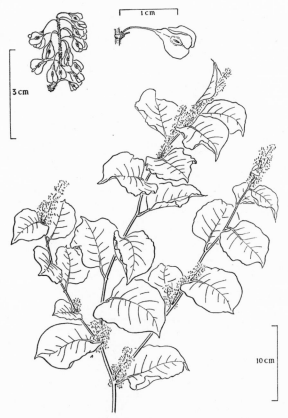

847.   *Polygonum cuspidatum* Sieb. & Zucc.   White

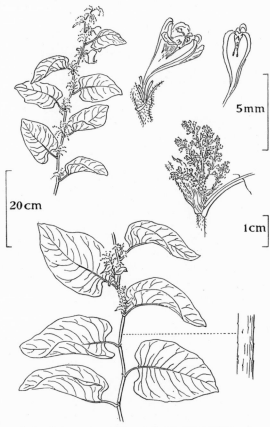

848.   *Polygonum sachalinense* F. Schmidt   White

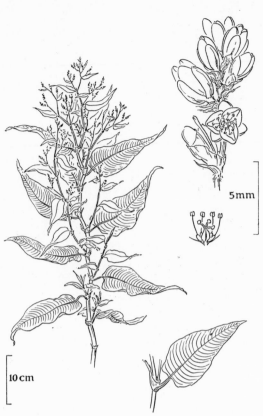

849. *Polygonum polystachyum* Wall. ex Meisn. White

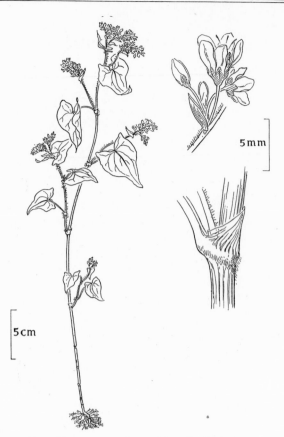

850. *Fagopyrum esculentum* Moench Buckwheat
      Pink or white

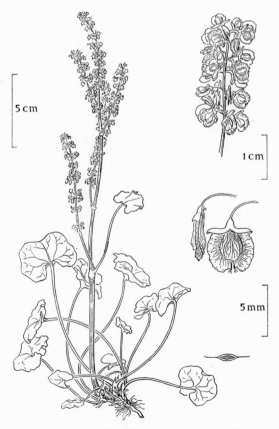

851. *Oxyria digyna* (L.) Hill Mountain Sorrel Greenish

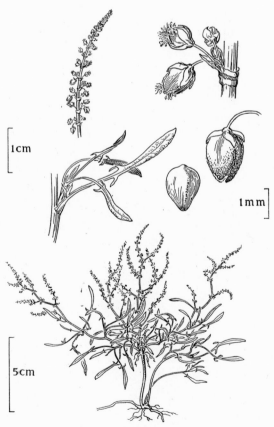

852. *Rumex tenuifolius* (Wallr.) Löve Sheep's Sorrel
      Reddish

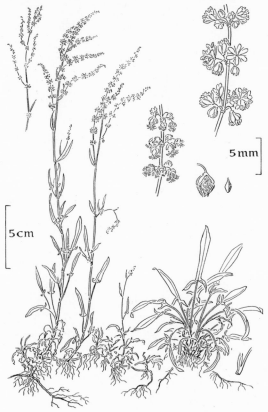

853.   *Rumex acetosella* L.   Sheep's Sorrel   Reddish

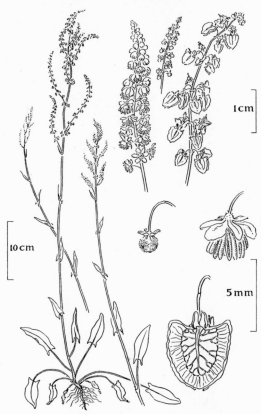

854.   *Rumex acetosa* L.   Sorrel   Reddish

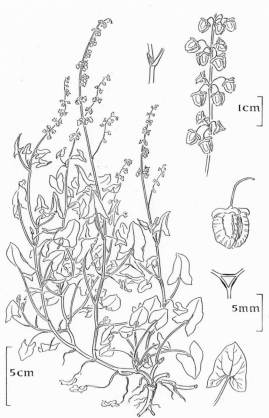

855.   *Rumex scutatus* L.   Reddish

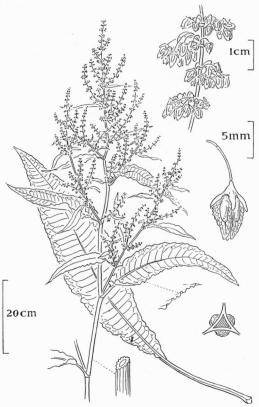

856.   *Rumex hydrolapathum* Huds.   Great Water Dock
Greenish

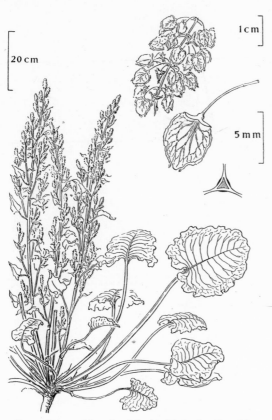

857. *Rumex alpinus* L. Monk's Rhubarb Greenish

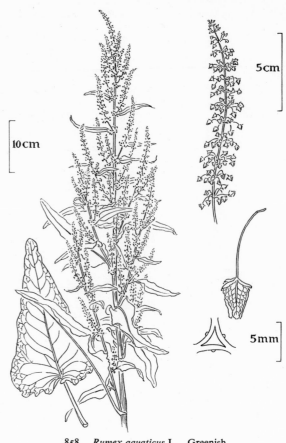

858. *Rumex aquaticus* L. Greenish

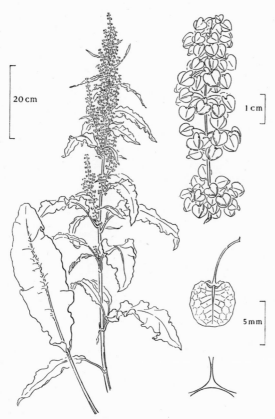

859. *Rumex longifolius* DC. Greenish

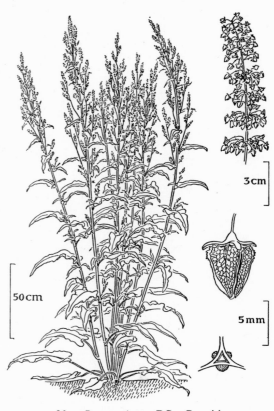

860. *Rumex cristatus* DC. Greenish

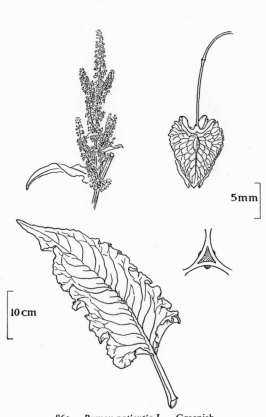

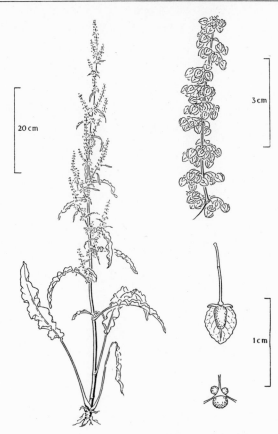

861.   *Rumex patientia* L.   Greenish

862.   *Rumex crispus* L.   Curled Dock   Greenish

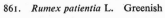

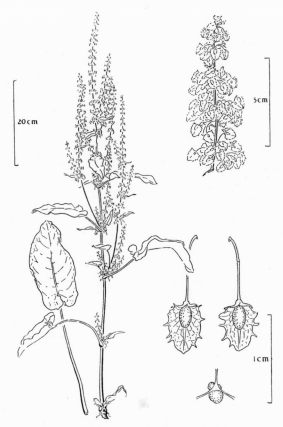

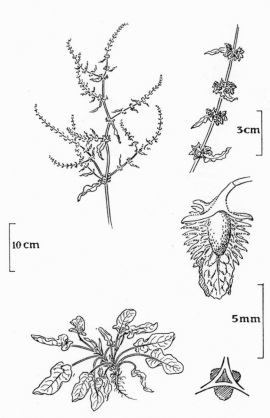

863.   *Rumex obtusifolius* L.   Broad-leaved Dock   Greenish

864.   *Rumex pulcher* L.   Fiddle Dock   Greenish

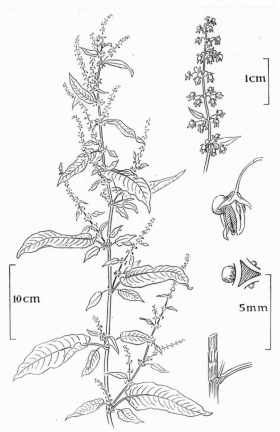

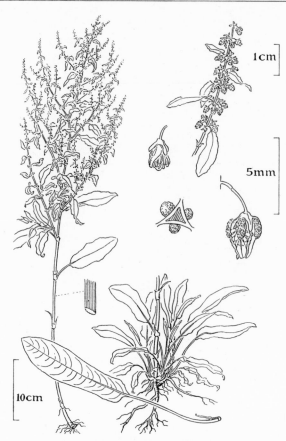

865. *Rumex sanguineus* L. Red-veined Dock Greenish

866. *Rumex conglomeratus* Murr. Sharp Dock Greenish

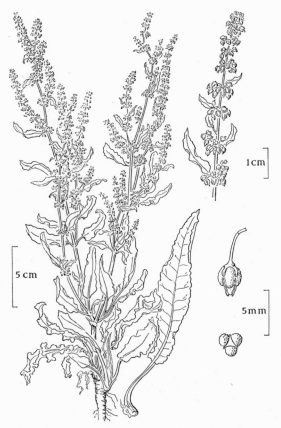

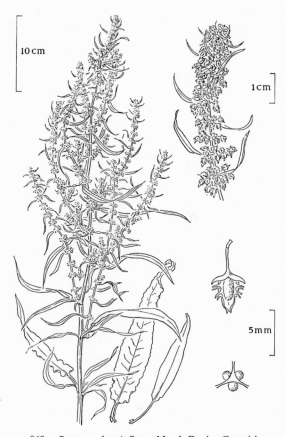

867. *Rumex rupestris* Le Gall Shore Dock Greenish

868. *Rumex palustris* Sm. Marsh Dock Greenish

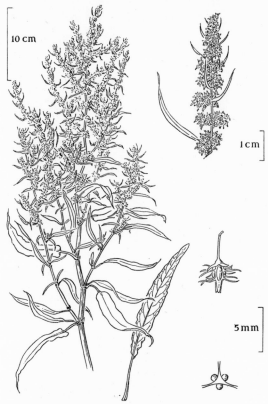

869. *Rumex maritimus* L.   Golden Dock   Greenish

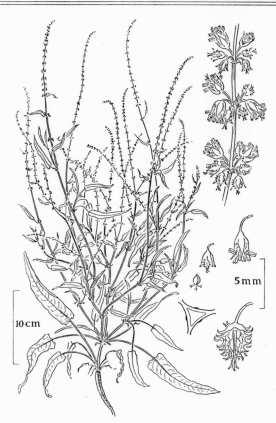

870. *Rumex brownii* Campd.   Greenish

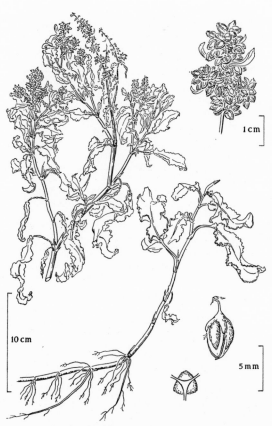

871. *Rumex frutescens* Thou.   Greenish

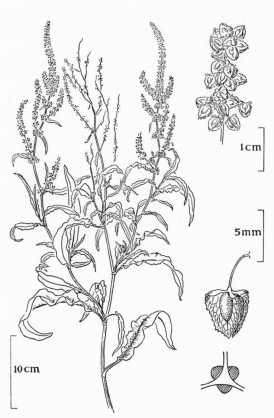

872. *Rumex triangulivalvis* (Danser) Rech. f.   Green

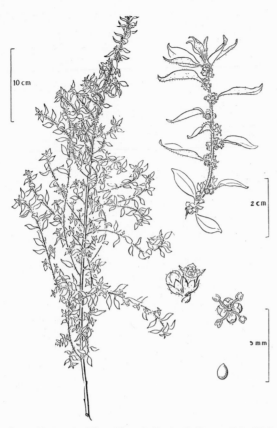

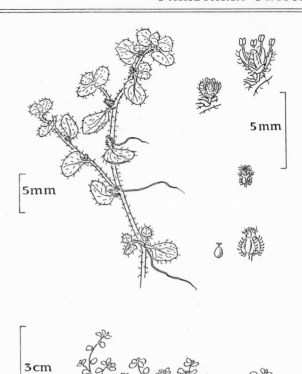

873. *Parietaria diffusa* Mert. & Koch   Pellitory-of-the-Wall
Green

874. *Helxine solierolii* Req.   Mind-your-own-business
Green

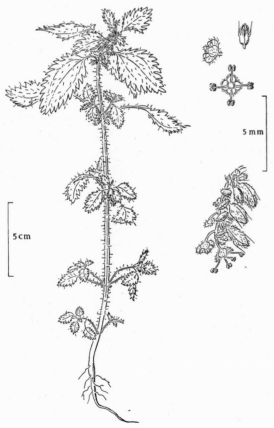

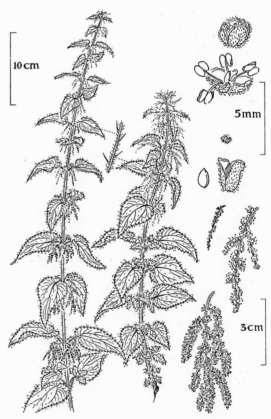

875. *Urtica urens* L. Small Nettle   Green

876. *Urtica dioica* L. Stinging Nettle   Green

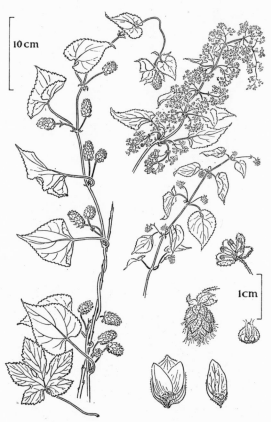

877. *Humulus lupulus* L. Hop Green

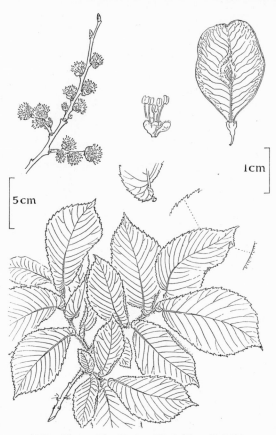

878. *Ulmus glabra* Huds. Wych Elm Reddish

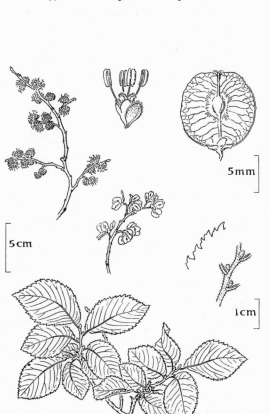

879. *Ulmus procera* Salisb. English Elm Reddish

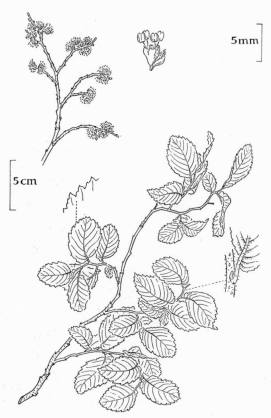

880. *Ulmus carpinifolia* Gleditsch Smooth Elm Reddish

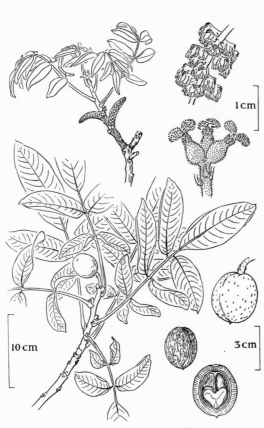

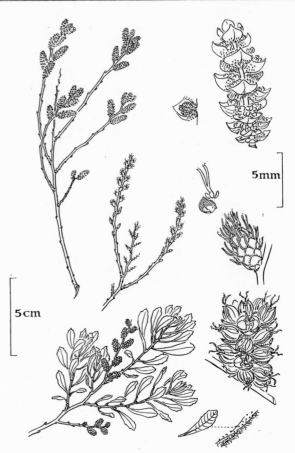

881. *Juglans regia* L.   Walnut   Green

882. *Myrica gale* L.   Bog Myrtle, Sweet Gale   Brown

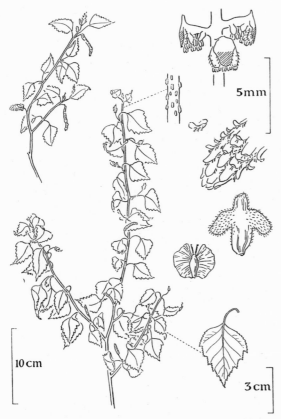

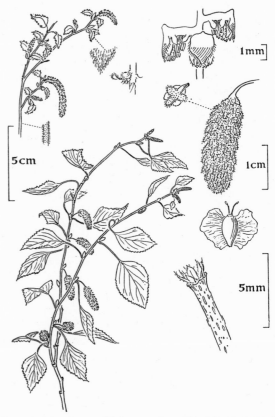

883. *Betula pendula* Roth.   Silver Birch   Greenish

884. *Betula pubescens* Ehrh. ssp. *odorata* (Bechst.) E. F. Warb.
Birch   Greenish

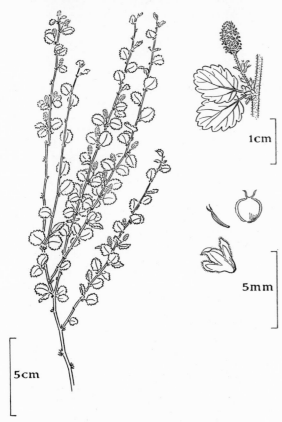

885.   *Betula nana* L.   Dwarf Birch   Greenish

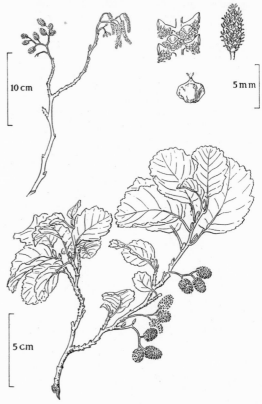

886.   *Alnus glutinosa* (L.) Gaertn.   Alder   Greenish

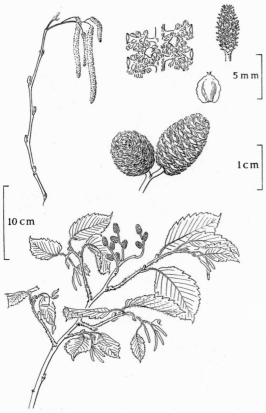

887.–*Alnus incana* (L.) Moench   Grey Alder   Greenish

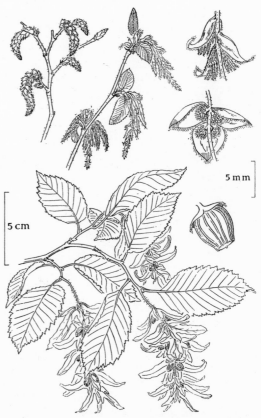

888.   *Carpinus betulus* L.   Hornbeam   Greenish

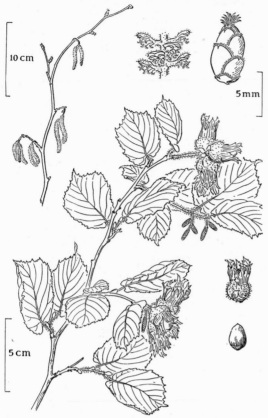

889. *Corylus avellana* L.   Hazel, Cob-nut   Greenish

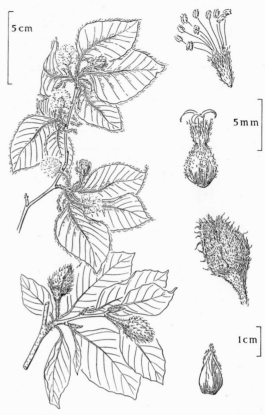

890. *Fagus sylvatica* L.   Beech   Greenish

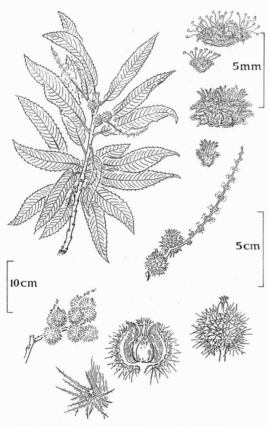

891. *Castanea sativa* L.   Sweet or Spanish Chestnut
Yellowish

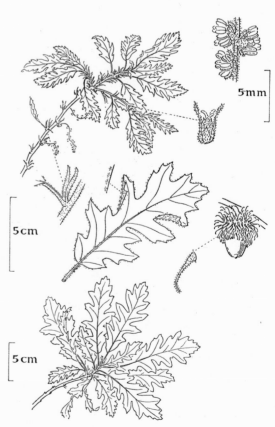

892. *Quercus cerris* L.   Turkey Oak   Greenish

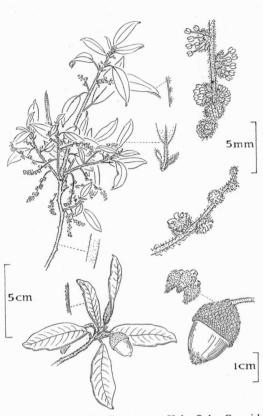

893. *Quercus ilex* L.   Evergreen or Holm Oak   Greenish

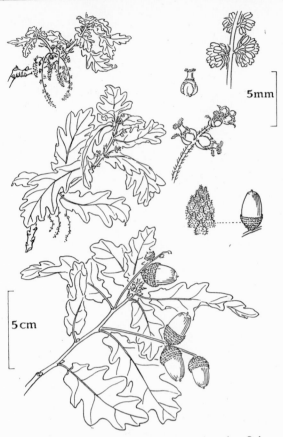

894. *Quercus robur* L.   Common or Pedunculate Oak
      Greenish

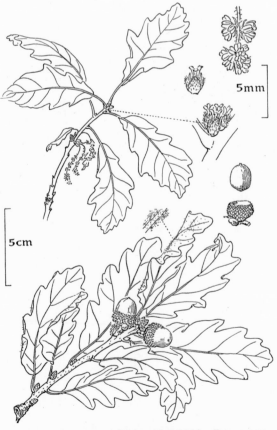

895. *Quercus petraea* (Mattuschka) Liebl.   Durmast or
      Sessile Oak   Greenish

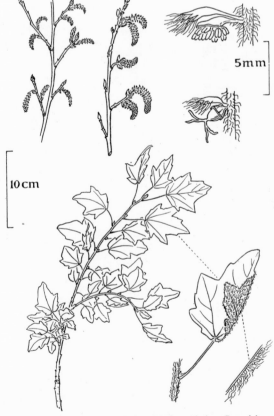

896. *Populus alba* L.   White Poplar, Abele   Greenish
      or purplish

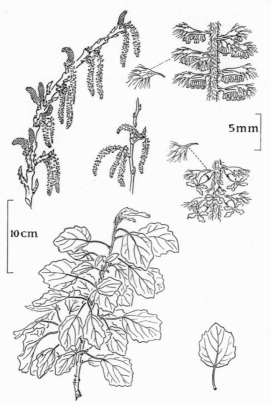

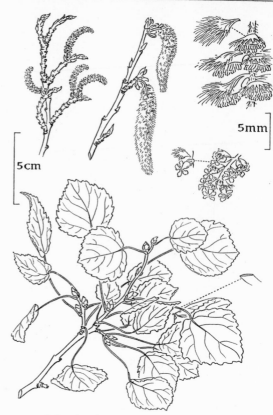

897. *Populus canescens* Sm.   Grey Poplar   Greenish
      or purplish

898. *Populus tremula* L.   Aspen   Purplish

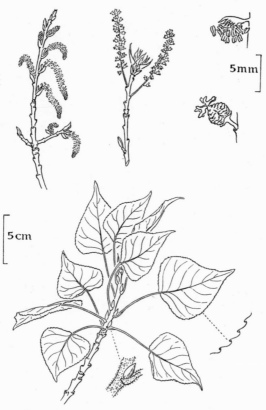

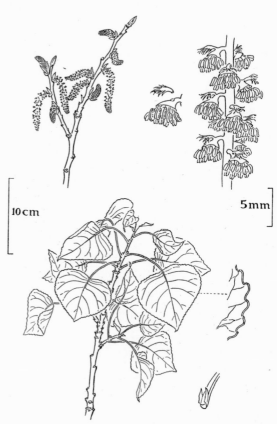

899. *Populus nigra* L.   Black Poplar   Reddish

900. *Populus × canadensis* Moench   Black Italian Poplar
      Reddish

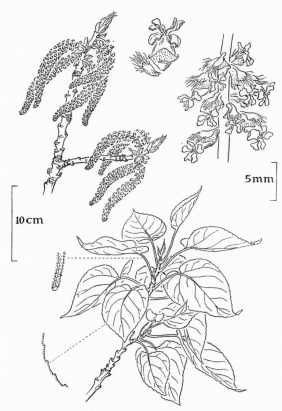

901. *Populus gileadensis* Rouleau Balm of Gilead Greenish

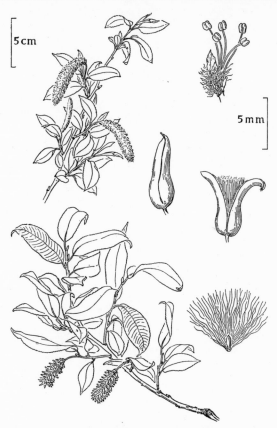

902. *Salix pentandra* L. Bay Willow Yellowish

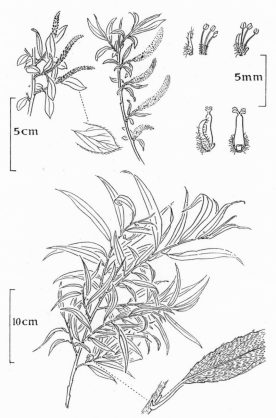

903. *Salix alba* L. White Willow Yellowish

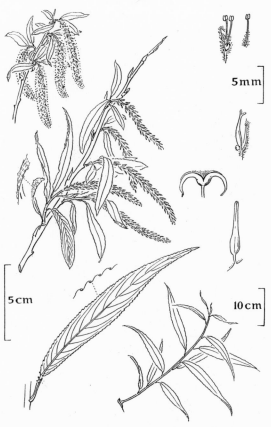

904. *Salix fragilis* L. Crack Willow Yellowish

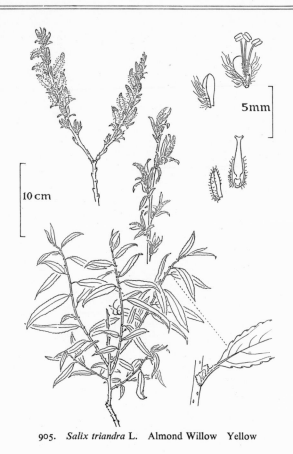

905.   *Salix triandra* L.   Almond Willow   Yellow

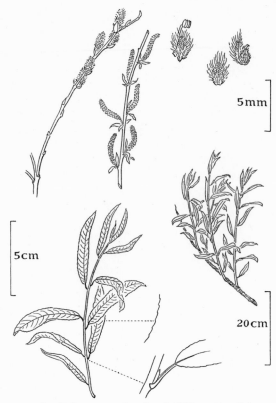

906.   *Salix purpurea* L.   Purple Willow   Purplish

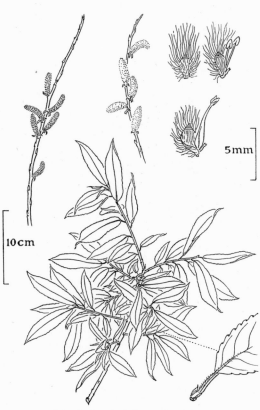

907.   *Salix daphnoides* Vill.   Yellowish or greyish

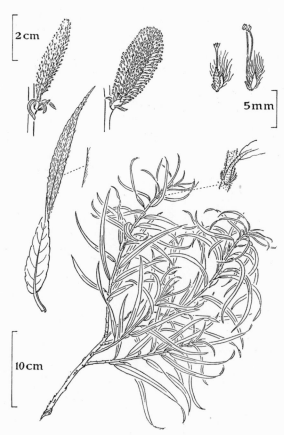

908.   *Salix viminalis* L.   Common Osier   Yellowish or greyish

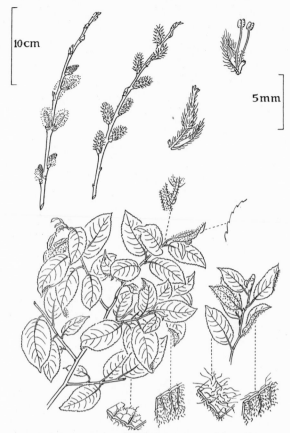

909. *Salix caprea* L.   Great Sallow, Goat Willow   Yellowish or greyish

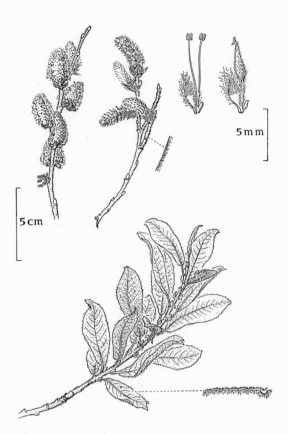

910. *Salix cinerea* L.   Common Sallow   Yellowish or greyish

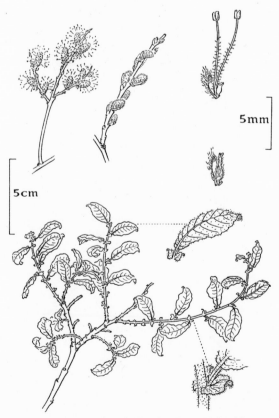

911. *Salix aurita* L.   Eared Sallow   Yellowish or greyish

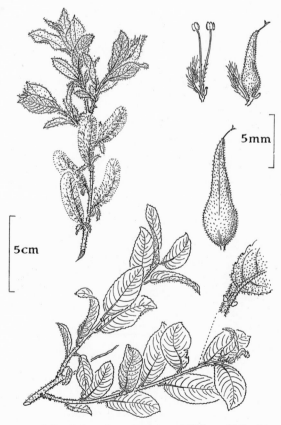

912. *Salix nigricans* Sm.   Dark-leaved Willow   Yellowish or greyish

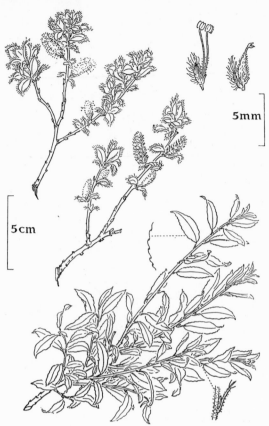

5mm

5cm

913. *Salix phylicifolia* L.   Tea-leaved Willow   Yellowish
            or greyish

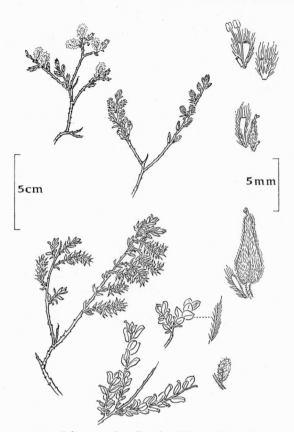

5cm

5mm

914. *Salix repens* L.   Creeping Willow   Yellowish
            or greyish

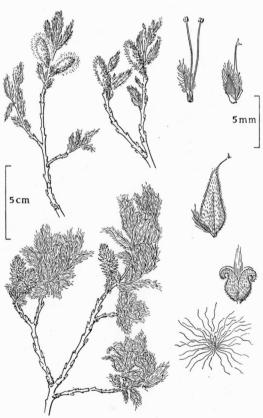

5mm

5cm

915. *Salix lapponum* L.   Downy Willow   Yellowish
            or greyish

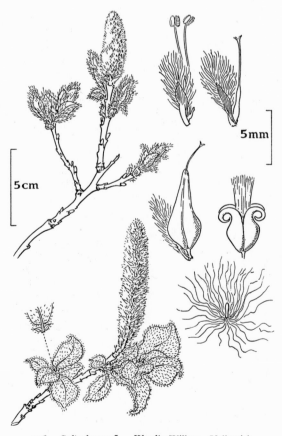

5mm

5cm

916. *Salix lanata* L.   Woolly Willow   Yellowish

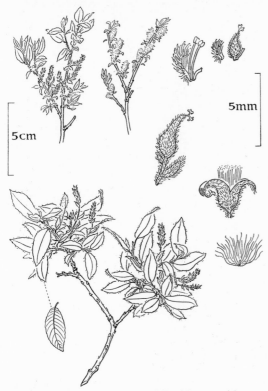

917.   *Salix arbuscula* L.   Yellowish or greyish

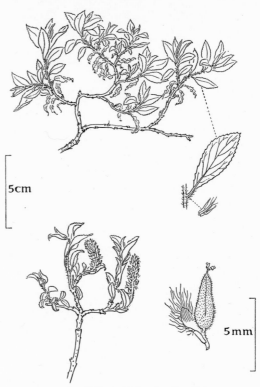

918.   *Salix myrsinites* L.   Purplish

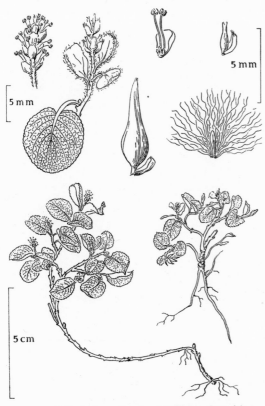

919.   *Salix herbacea* L.   Least Willow   Greenish

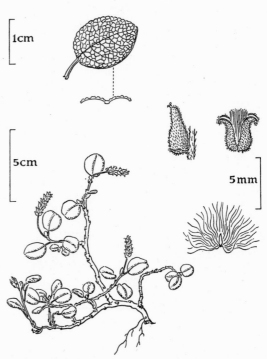

920.   *Salix reticulata* L.   Reticulate Willow   Greyish

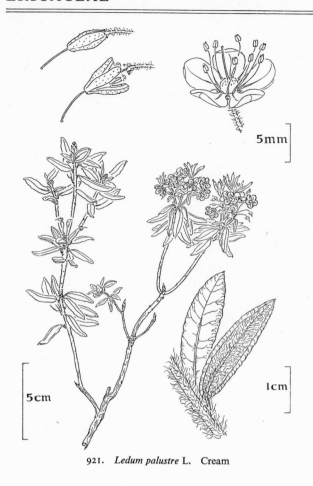

921.   *Ledum palustre* L.   Cream

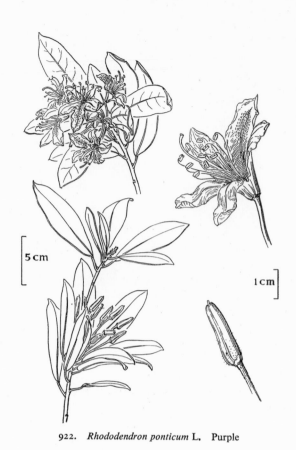

922.   *Rhododendron ponticum* L.   Purple

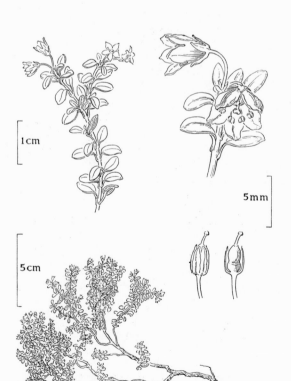

923.   *Loiseleuria procumbens* (L.) Desv.   Pink

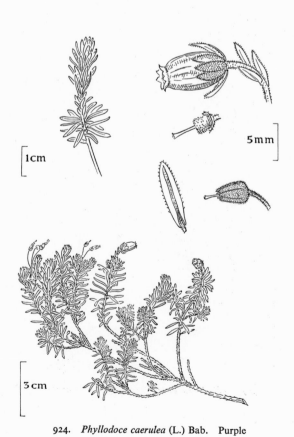

924.   *Phyllodoce caerulea* (L.) Bab.   Purple

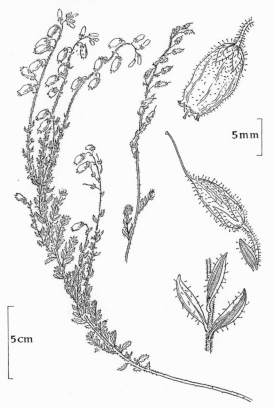

925. *Daboecia cantabrica* (Huds.) C. Koch   St Dabeoc's
          Heath   Purple

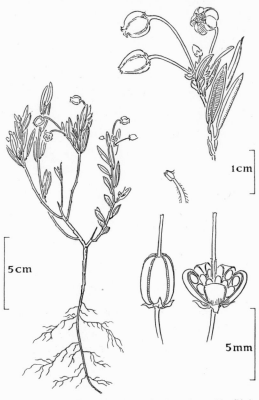

926. *Andromeda polifolia* L.   Marsh Andromeda   Pink

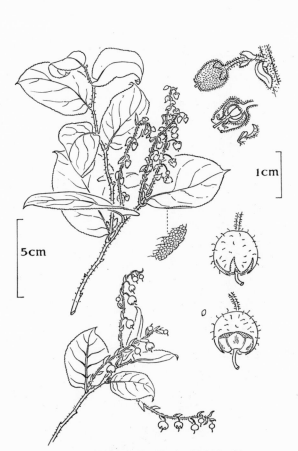

927. *Gaultheria shallon* Pursh   Pinkish-white

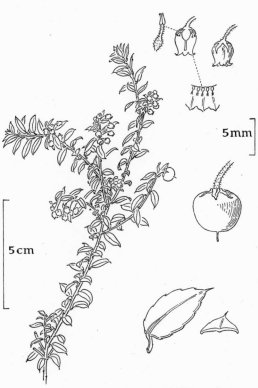

928. *Pernettya mucronata* (L. f.) Gaudich. ex Spreng.
          White

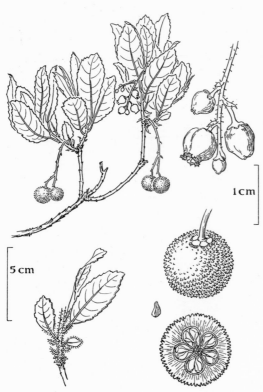

929. *Arbutus unedo* L. Strawberry Tree White

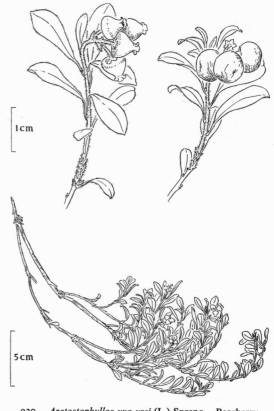

930. *Arctostaphyllos uva-ursi* (L.) Spreng. Bearberry
Pinkish-white

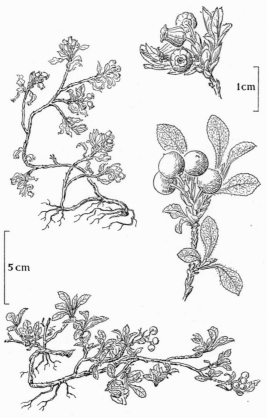

931. *Arctous alpinus* (L.) Nied. Black Bearberry White

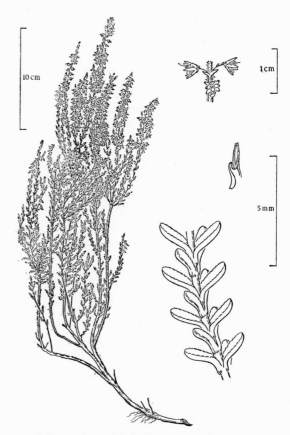

932. *Calluna vulgaris* (L.) Hull Ling, Heather Pale purple

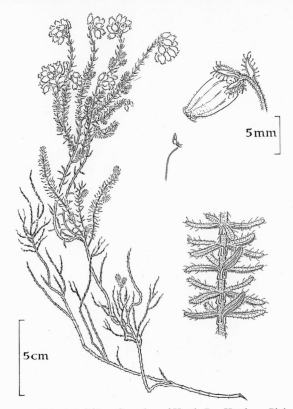

933.   *Erica tetralix* L.   Cross-leaved Heath, Bog Heather   Pink

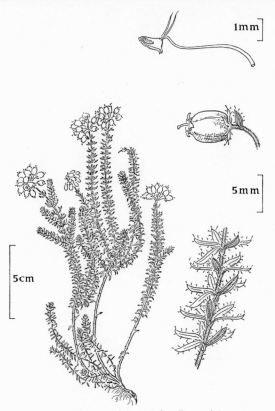

934.   *Erica mackaiana* Bab.   Deep pink

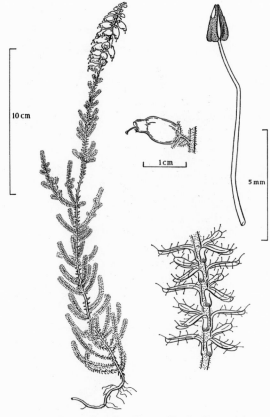

935.   *Erica ciliaris* L.   Dorset Heath   Deep pink

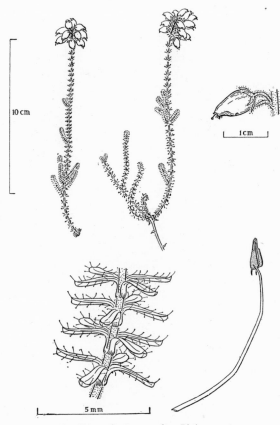

936.   *Erica ciliaris × tetralix*   Pink

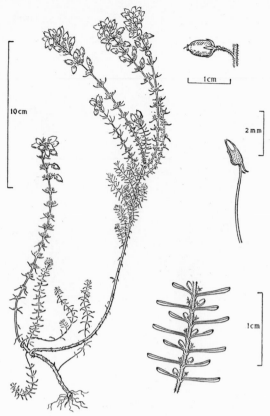

937. *Erica cinerea* L. Bell-heather Purple

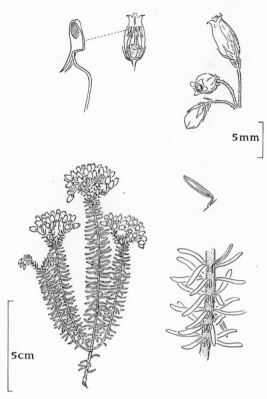

938. *Erica terminalis* Salisb. Pink

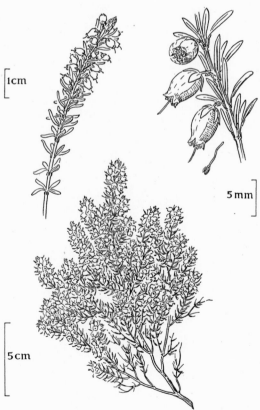

939. *Erica mediterranea* L. Irish Heath Purplish-pink

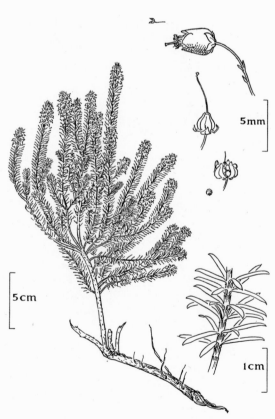

940. *Erica vagans* L. Cornish Heath Pale lilac

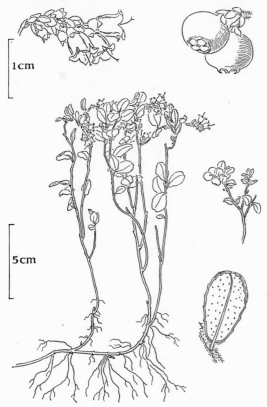

941.   *Vaccinium vitis-idaea* L. Cowberry Pinkish-white

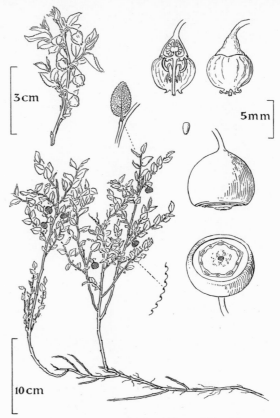

942.   *Vaccinium myrtillus* L. Bilberry, Whortleberry
Greenish-pink

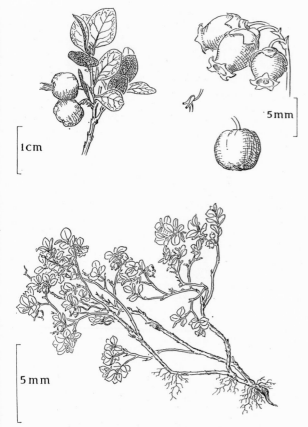

943.   *Vaccinium uliginosum* L. Bog Whortleberry Pale pink

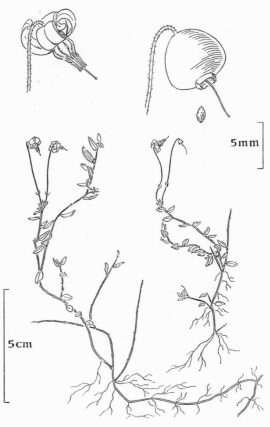

944.   *Vaccinium oxycoccus* L. Cranberry Pink

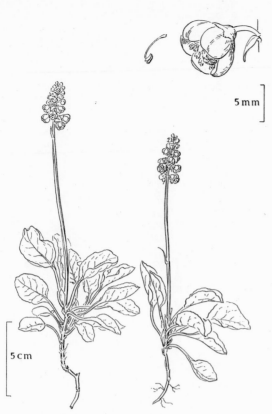

945. *Pyrola minor* L.   Common Wintergreen   Pinkish

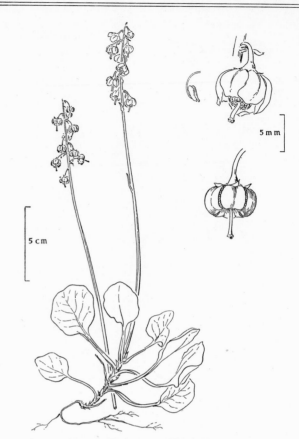

946. *Pyrola media* Sw.   Intermediate Wintergreen   Whitish

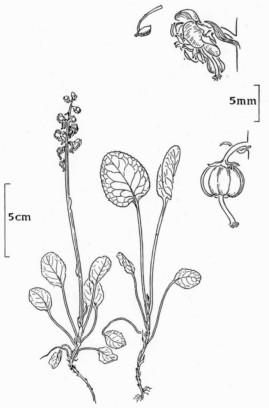

947. *Pyrola rotundifolia* L.   ssp. *rotundifolia*   Larger Winter-
       green   White

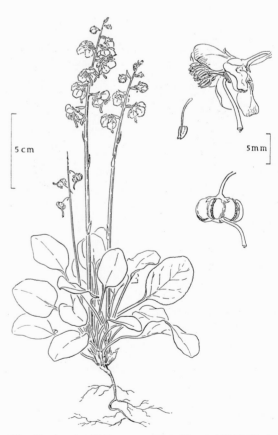

948. *Pyrola rotundifolia* ssp. *maritima* (Kenyon) E. F. Warb.
       White

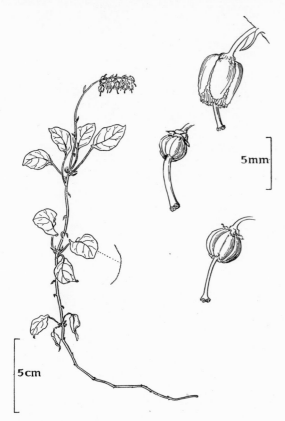

949. *Orthilia secunda* (L.) House   Serrated Wintergreen
            Greenish-white

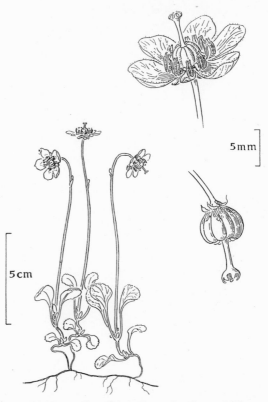

950. *Moneses uniflora* (L.) A. Gray   One-flowered Winter-
            green   White

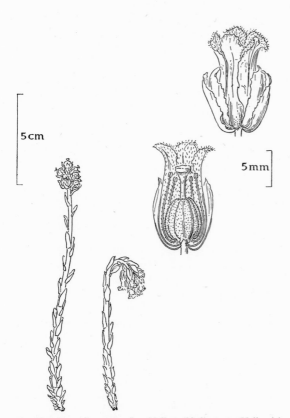

951. *Monotropa hypopitys* L.   Yellow Bird's-nest   Yellowish

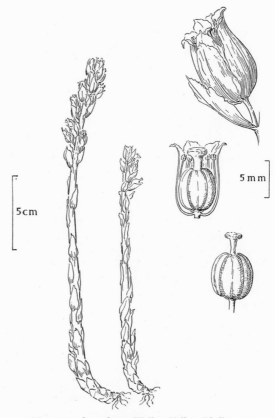

952. *Monotropa hypophegea* Wallr. Yellow Bird's-nest
            Yellowish

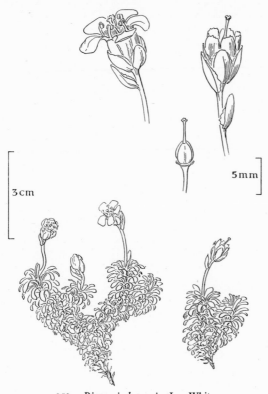

953. *Diapensia lapponica* L.   White

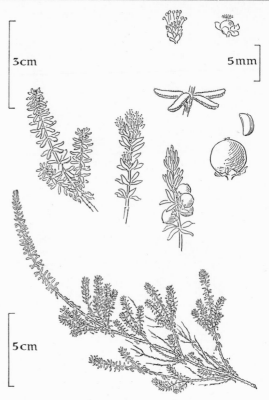

954. *Empetrum nigrum* L.   Crowberry   Pinkish

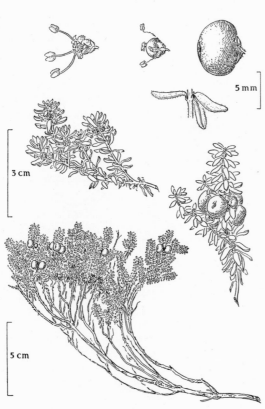

955. *Empetrum hermaphroditum* Hagerup   Crowberry
Pinkish

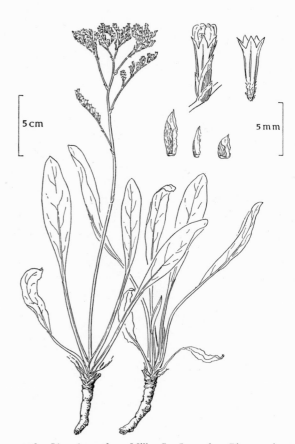

956. *Limonium vulgare* Mill.   Sea Lavender   Blue-purple

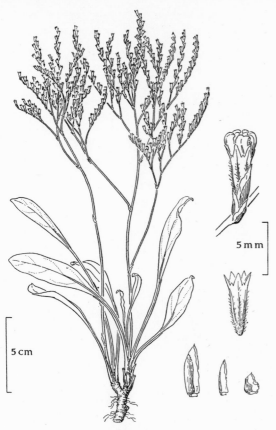

957. *Limonium humile* Mill.   Lax-flowered Sea Lavender
Blue-purple

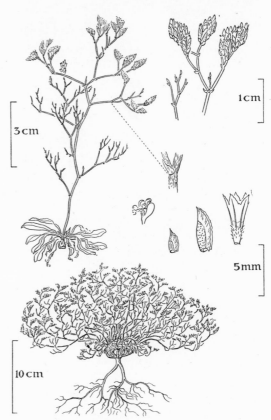

958. *Limonium bellidifolium* (Gouan) Dum.   Matted
Sea Lavender   Pale lilac

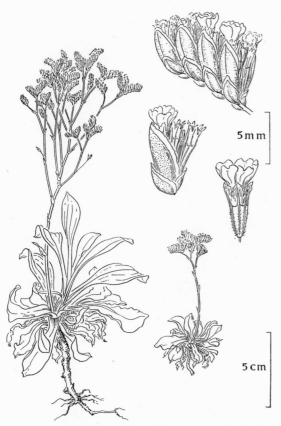

959. *Limonium auriculae-ursifolium* (Pourr.) Druce
Violet-blue

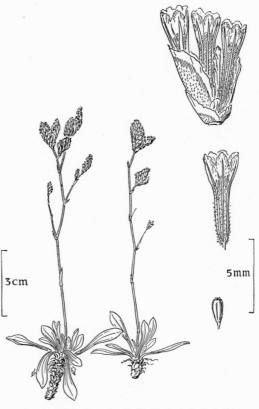

960. *Limonium binervosum* (G. E. Sm.) C. E. Salmon
Rock Sea Lavender   Violet-blue

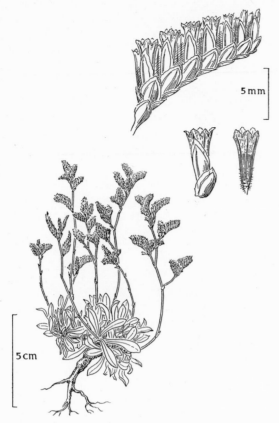

961.   *Limonium recurvum* C. E. Salmon   Violet-blue

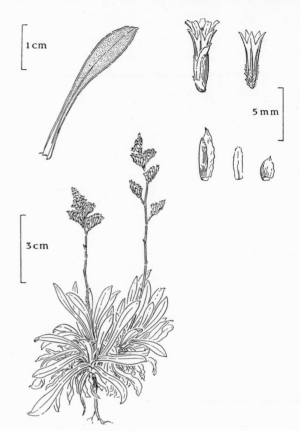

962.   *Limonium transwallianum* (Pugsl.) Pugsl.   Violet-blue

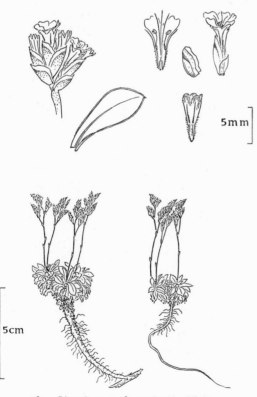

963.   *Limonium paradoxum* Pugsl.   Violet

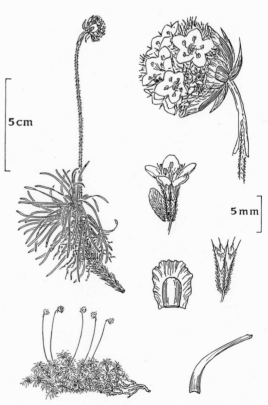

964.   *Armeria maritima* (Mill.) Willd.   Thrift, Sea Pink
Pink or white

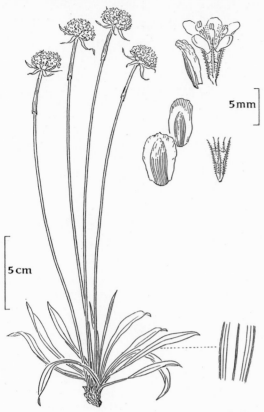

965. *Armeria arenaria* (Pers.) Schult. Jersey Thrift
Deep pink

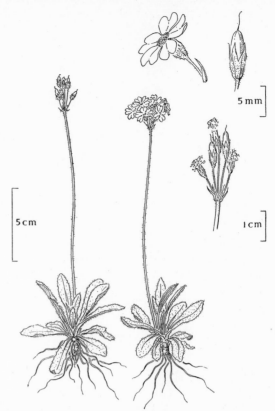

966. *Primula farinosa* L. Bird's-eye Primrose Rosy lilac

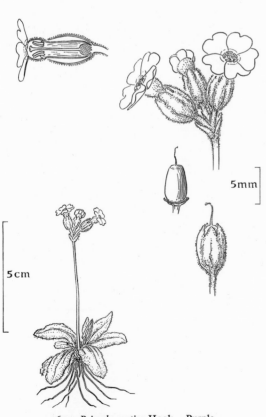

967. *Primula scotica* Hook. Purple

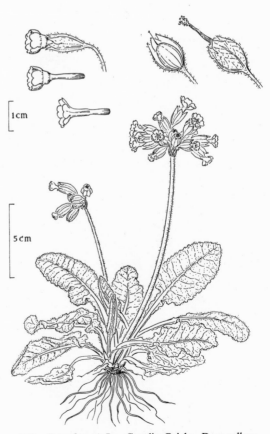

968. *Primula veris* L. Cowslip, Paigle Deep yellow

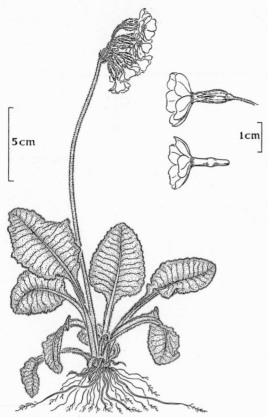

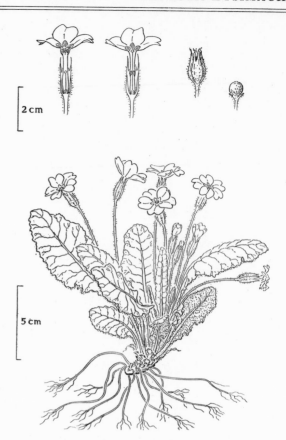

969. *Primula elatior* (L.) Hill   Oxlip, Paigle   Pale yellow

970. *Primula vulgaris* Huds.   Primrose   Pale yellow

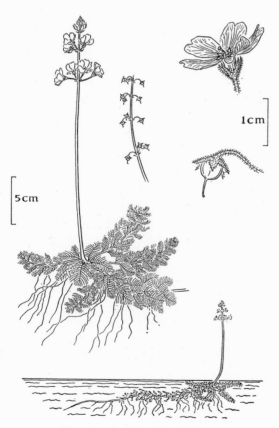

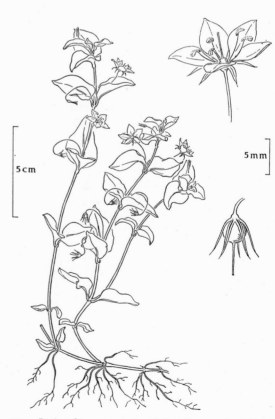

971. *Hottonia palustris* L.   Water Violet   Lilac

972. *Lysimachia nemorum* L.   Yellow Pimpernel   Yellow

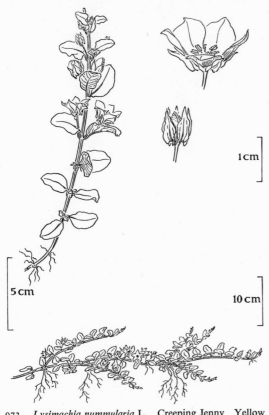

973. *Lysimachia nummularia* L.   Creeping Jenny   Yellow

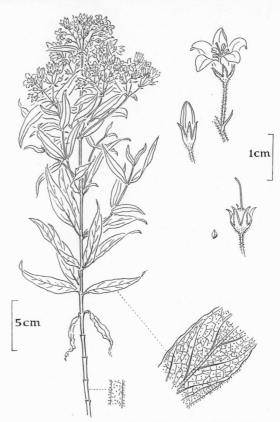

974. *Lysimachia vulgaris* L.   Yellow Loosestrife   Yellow

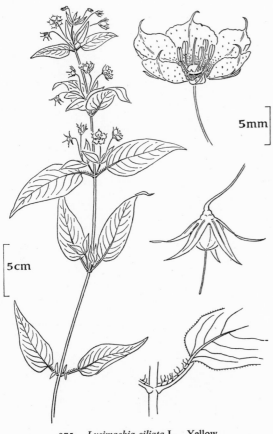

975. *Lysimachia ciliata* L.   Yellow

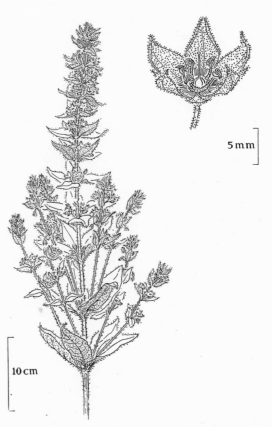

976. *Lysimachia punctata* L.   Yellow

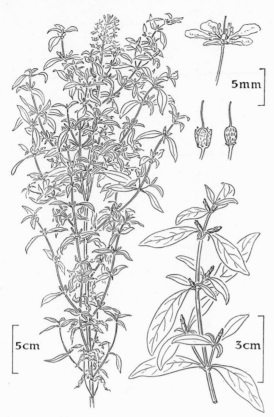

977. *Lysimachia terrestris* (L.) Britton, Sterns & Poggenb.
Yellow

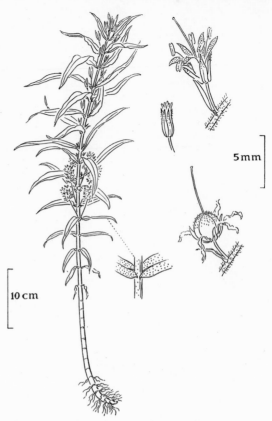

978. *Naumburgia thyrsiflora* (L.) Rchb.   Tufted Loosestrife
Yellow

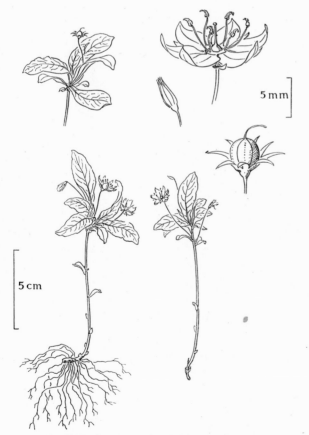

979. *Trientalis europaea* L.   Chickweed Wintergreen
White

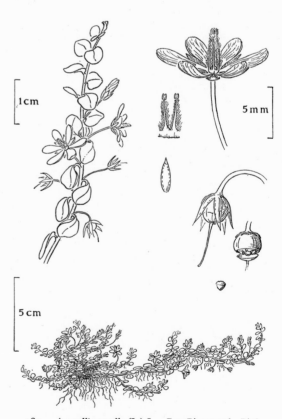

980. *Anagallis tenella* (L.) L.   Bog Pimpernel   Pink

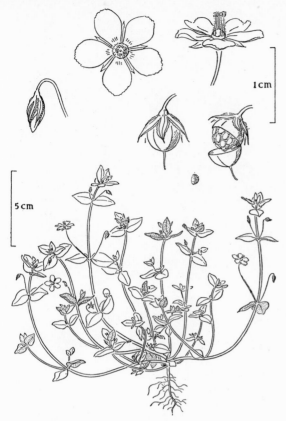

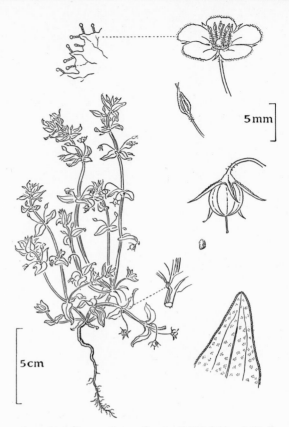

981. *Anagallis arvensis* L. ssp. *arvensis* Scarlet Pimpernel
Red (blue)

982. *Anagallis arvensis* ssp. *foemina* (Mill.) Schinz & Thell.
Blue

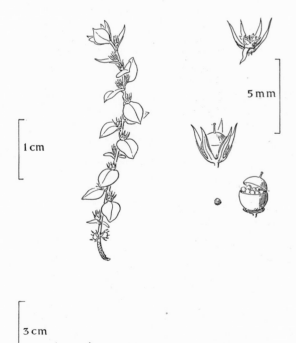

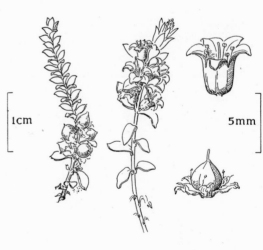

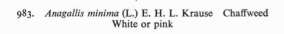

983. *Anagallis minima* (L.) E. H. L. Krause Chaffweed
White or pink

984. *Glaux maritima* L. Sea Milkwort, Black Saltwort
Pink

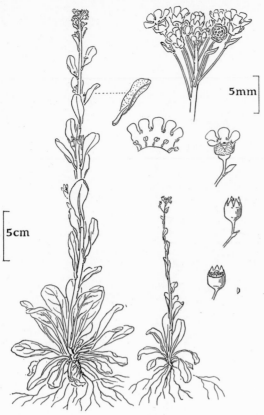

985. *Samolus valerandi* L.   Brookweed   White

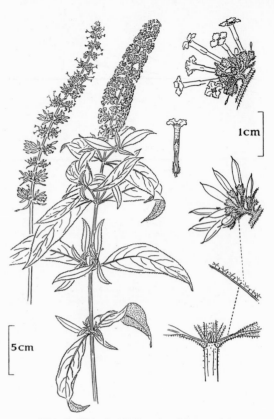

986. *Buddleja davidii* Franch.   Lilac or violet

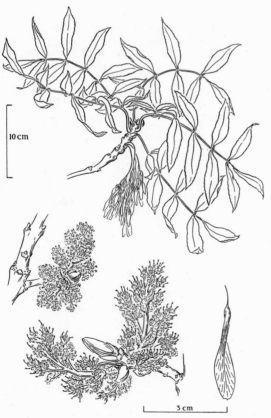

987. *Fraxinus excelsior* L.   Ash   Purplish

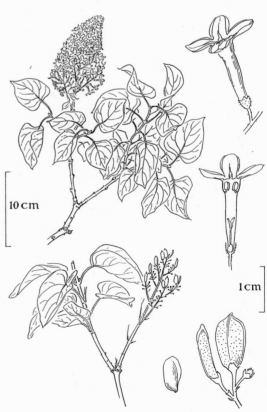

988. *Syringa vulgaris* L.   Lilac   Lilac or white

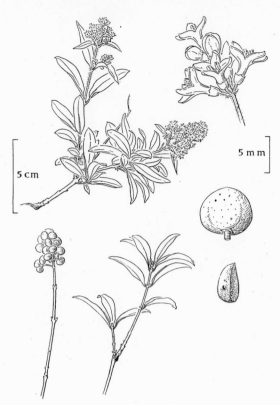

989. *Ligustrum vulgare* L. Common Privet White

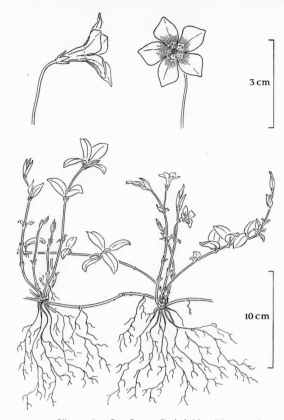

990. *Vinca minor* L. Lesser Periwinkle Blue-purple

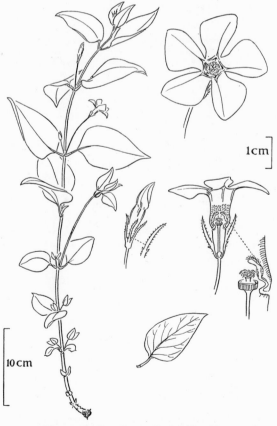

991. *Vinca major* L. Greater Periwinkle Blue-purple

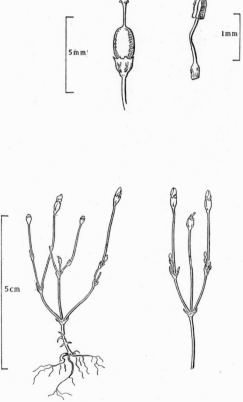

992. *Cicendia filiformis* (L.) Delarb. Yellow

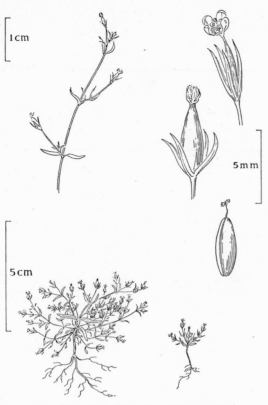

1cm

5mm

5cm

993. *Exaculum pusillum* (Lam.) Caruel   Pink

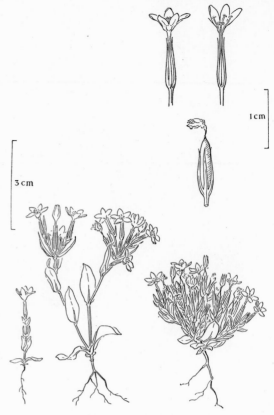

1cm

3cm

994. *Centaurium pulchellum* (Sw.) Druce   Pink

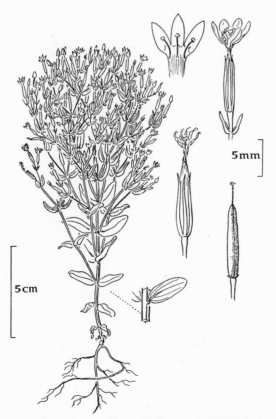

5mm

5cm

995. *Centaurium tenuiflorum* (Hoffmanns. & Link) Fritsch
Pink

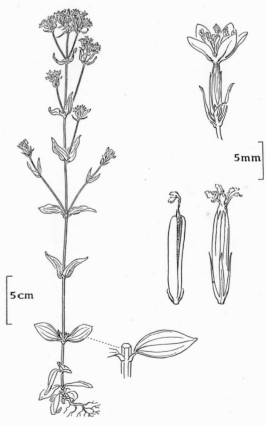

5mm

5cm

996. *Centaurium erythraea* Rafn   Common Centaury
Pink

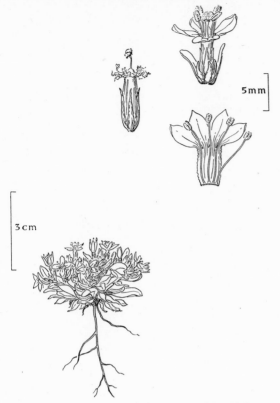

997.   *Centaurium capitatum* (Willd.) Borbás   Pink

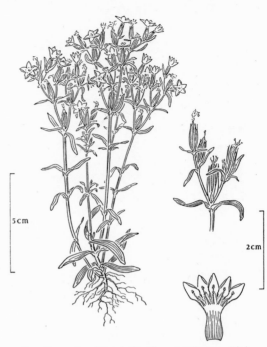

998.   *Centaurium littorale* (D. Turner) Gilmour   Pink

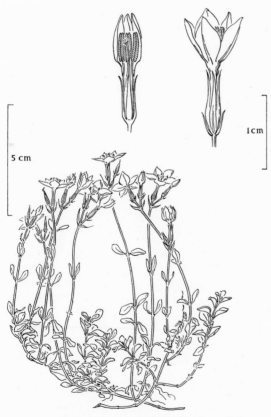

999.   *Centaurium portense* (Brot.) Butcher   Pink

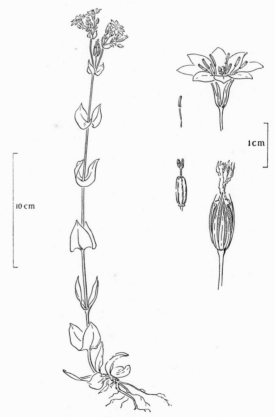

1000.   *Blackstonia perfoliata* (L.) Huds.   Yellow-wort
Yellow

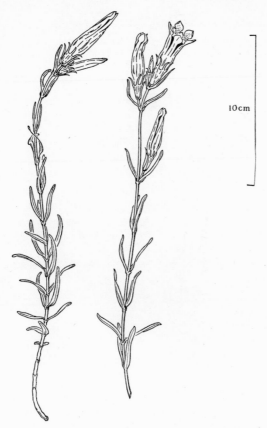

1001.   *Gentiana pneumonanthe* L.   Marsh Gentian   Blue

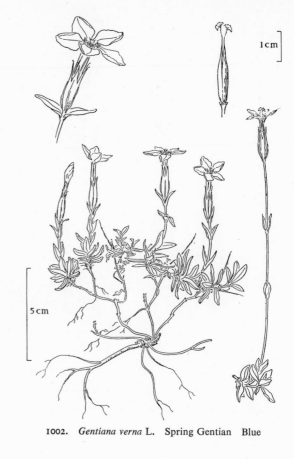

1002.   *Gentiana verna* L.   Spring Gentian   Blue

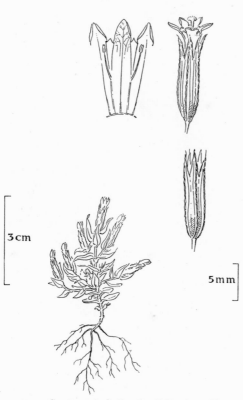

1003.   *Gentiana nivalis* L.   Small Gentian   Blue

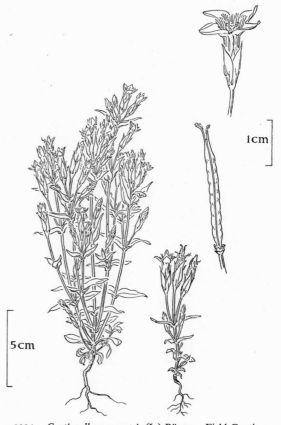

1004.   *Gentianella campestris* (L.) Börner   Field Gentian
        Bluish-lilac

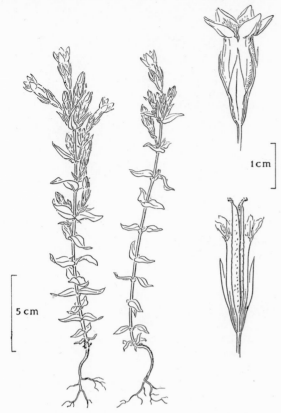

1005.   *Gentianella germanica* (Willd.) Börner   Lilac

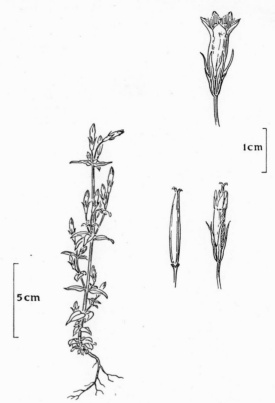

1006.   *Gentianella amarella* (L.) Börner   Felwort
Dull purple

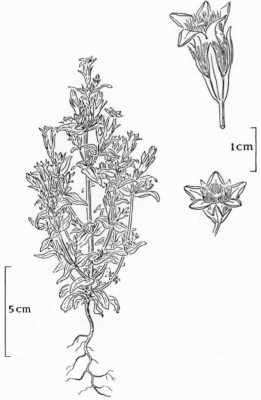

1007.   *Gentianella amarella* ssp. *septentrionalis* (Druce)
Pritchard   Cream and reddish-purple

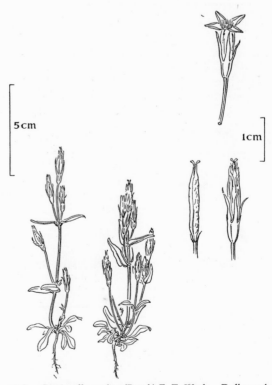

1008.   *Gentianella anglica* (Pugsl.) E. F. Warb.   Dull purple

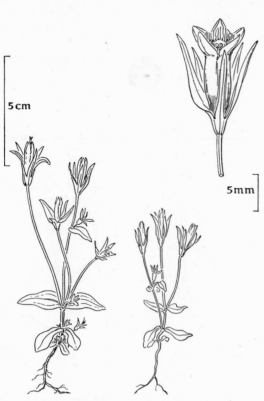

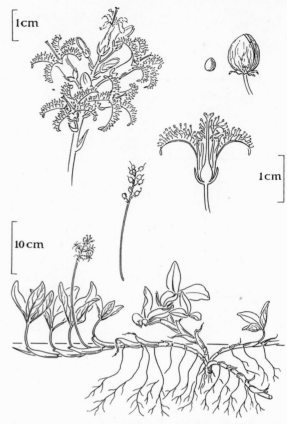

1009. *Gentianella uliginosa* (Willd.) Börner Dull purple

1010. *Menyanthes trifoliata* L. Bogbean, Buckbean
Pinkish-white

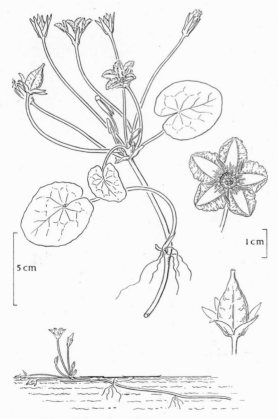

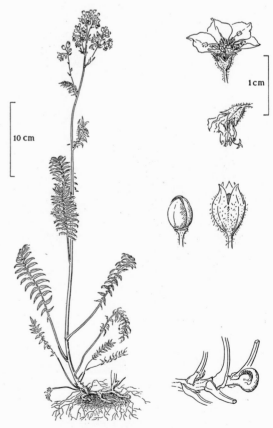

1011. *Nymphoides peltata* (S. G. Gmel.) Kuntze Yellow

1012. *Polemonium caeruleum* L. Jacob's Ladder Blue

INDEX

Erica (cont.)
  terminalis Salisb., 938
  tetralix L., 933
  vagans L., 940
Eryngium campestre L., 742
  maritimum L., 741
Euphorbia amygdaloides L., 827
  corallioides L., 814
  cyparissias L., 826
  dulcis L., 816
  esula L., 825
  exigua L., 821
  helioscopia L., 819
  hyberna L., 815
  lathyrus L., 812
  paralias L., 823
  peplis L., 811
  peplus L., 820
  pilosa L., 813
  platyphyllos L., 817
  portlandica L., 822
  stricta L., 818
  uralensis Fisch., 824
Evening Primrose, 717–19
Exaculum pusillum (Lam.) Caruel, 993

Fagopyrum esculentum Moench, 850
Fagus sylvatica L., 890
Falcaria vulgaris Bernh., 771
Felwort, 1006
Fennel, 791
Filipendula ulmaria (L.) Maxim., 555
  vulgaris Moench, 554
Fireweed, 716
Foeniculum vulgare Mill., 791
Fool's Parsley, 790
Fool's Watercress, 765
Fragaria ananassa Duchesne, 579
  moschata Duchesne, 578
  vesca L., 577
Fraxinus excelsior L., 987
Fuchsia magellanica Lam., 720

Gaultheria shallon Pursh, 927
Gean, 621
Gentian, 1001–4
Gentiana nivalis L., 1003
  pneumonanthe L., 1001
  verna L., 1002
Gentianella amarella (L.) Börner, 1006–7
  anglica (Pugsl.) E. F. Warb., 1008
  campestris (L.) Börner, 1004
  germanica (Willd.) Börner, 1005
  uliginosa (Willd.) Börner, 1009
Geum rivale L., 581
  urbanum L., 580
Glaux maritima L., 984
Golden Saxifrage, 683–4
Gooseberry, 691
Goutweed, 778
Grass of Parnassus, 685
Grass Poly, 697
Ground Elder, 778

Hare's-ear, 759, 761
Hawthorn, 628–9
Hazel, 889
Heath, Cornish, 940
  Cross-leaved, 933
  Dorset, 935
  Irish, 939
Heather, 932
  Bog, 933
Hedera helix L., 737
Hedge-parsley, 750–2
Helxine solierolii Req., 874
Hemlock, 757
Heracleum mantegazzianum Somm. & Lev., 803
  sphondylium L., 802
Herb Bennet, 580
Hippophaë rhamnoides L., 701
Hippuris vulgaris L., 726
Hog's Fennel, 798–9

Hogweed, 802
Honewort, 763
Hop, 877
Hornbeam, 888
Hottonia palustris L., 971
Houseleek, 665
Humulus lupulus L., 877
Hydrocotyle vulgaris L., 738

Ivy, 737

Jacob's Ladder, 1012
Juglans regia L., 881

Keck, 746, 802
Knotgrass, 829, 831
  Ray's, 832
  Sea, 833
Koenigia islandica L., 828

Lady's-Mantle, 586–98
Ledum palustre L., 921
Ligusticum scoticum L., 795
Ligustrum vulgare L., 989
Lilac, 988
Limonium auriculae-ursifolium (Pourr.) Druce, 959
  bellidifolium (Gouan) Dum., 958
  binervosum (G. E. Sm.) C. E. Salmon, 960
  humile Mill., 957
  paradoxum Pugsl., 963
  recurvum C. E. Salmon, 961
  transwallianum (Pugsl.) Pugsl., 962
  vulgare Mill., 956
Ling, 932
Livelong, 655
Loiseleuria procumbens (L.) Desv., 923
London Pride, 671
Loosestrife, Purple, 696
  Tufted, 978
  Yellow, 974
Lovage, 795
Ludwigia palustris (L.) Elliott, 702
Lysimachia ciliata L., 975
  nemorum L., 972
  nummularia L., 973
  punctata L., 976
  terrestris (L.) Britton, Sterns & Poggenb., 977
  vulgaris L., 974
Lythrum hyssopifolia L., 697
  salicaria L., 696

Malus sylvestris Mill., 653
Mare's-tail, 726
Masterwort, 800
Meadow-sweet, 555
Medlar, 630
Menyanthes trifoliata L., 1010
Mercurialis annua L., 810
  perennis L., 809
Mespilus germanica L., 630
Meu, 793
Meum athamanticum Jacq., 793
Mezereon, 699
Midsummer-men, 654
Milk Parsley, 799
Mind-your-own-business, 874
Mistletoe, 733
Mock Orange, 686
Moneses uniflora (L.) A. Gray, 950
Monk's Rhubarb, 857
Monotropa hypophegea Wallr., 952
  hypopitys L., 951
Mountain Ash, 632
Myrica gale L., 882
Myriophyllum alterniflorum DC., 725
  spicatum L., 724
  verticillatum L., 723
Myrrhis odorata (L.) Scop., 749

Naumburgia thyrsiflora (L.) Rchb., 978
Navelwort, 667

Nettle, Small, 875
  Stinging, 876
Nymphoides peltata (S. G. Gmel.) Kuntze, 1011

Oak, 892–5
Oenanthe aquatica (L.) Poir., 788
  crocata L., 787
  fistulosa L., 783
  fluviatilis (Bab.) Colem., 789
  lachenalii C. C. Gmel., 786
  pimpinelloides L., 784
  silaifolia M. Bieb., 785
Oenothera biennis L., 717
  erythrosepala Borbás, 718
  stricta Ledeb., 719
Orpine, 655
Orthilia secunda (L.) House, 949
Osier, 908
Oxlip, 969
Oxyria digyna (L.) Hill, 851

Paigle, 968–9
Parietaria diffusa Mert. & Koch, 873
Parnassia palustris L., 685
Parsley, 767
Parsley Piert, 599
Parsnip, Wild, 801
Pastinaca sativa L., 801
Pear, 652
Pellitory-of-the-Wall, 873
Pennywort, 738
Peplis portula L., 698
Pepper Saxifrage, 792
Periwinkle, 990–1
Pernettya mucronata (L. f.) Gaudich., 928
Persicaria, 838–9
Petroselinum crispum (Mill.) Nym., 767
  segetum (L.) Koch, 768
Peucedanum officinale L., 798
  ostruthium (L.) Koch, 800
  palustre (L.) Moench, 799
Philadelphus coronarius L., 686
Phyllodoce caerulea (L.) Bab., 924
Physospermum cornubiense (L.) DC., 756
Pimpernel, Bog, 980
  Scarlet, 981
  Yellow, 972
Pimpinella major (L.) Huds., 777
  saxifraga L., 776
Pitcher-plant, 695
Polemonium caeruleum L., 1012
Polygonum amphibium L., 836–7
  arenastrum Bor., 831
  aviculare L., 829
  bistorta L., 835
  convolvulus L., 844
  cuspidatum Sieb. & Zucc., 847
  dumetorum L., 845
  hydropiper L., 841
  lapathifolium L., 839
  maritimum L., 833
  minus Huds., 843
  mite Schrank, 842
  nodosum Pers., 840
  persicaria L., 838
  polystachyum Wall., 849
  raii Bab., 832
  rurivagum Jord., 830
  sachalinense F. Schmidt, 848
  sagittatum L., 846
  viviparum L., 834
Populus alba L., 896
  × *canadensis* Moench, 900
  canescens Sm., 897
  gileadensis Rouleau, 901
  nigra L., 899
  tremula L., 898
Potentilla anglica Laicharding, 574
  anserina L., 566
  argentea L., 567
  crantzii (Crantz) G. Beck, 572
  erecta (L.) Räusch., 573